Sustainable Finance

Series Editors

Sustainable Finance is a concise and authoritative reference series linking research and practice. It provides reliable concepts and research findings in the ever growing field of sustainable investing and finance, SDG economics and Leadership with the declared commitment to present the theories, methods, tools and investment approaches that can fulfil the United Nations Sustainable Development Goals and the Paris Agreement COP 21/22 alongside with de-risking assets and creating triple purpose solutions that ensure the parity of profit, people and planet through choice architecture passion and performance. The series addresses market failure, systemic risk and reinvents portfolio theory, portfolio engineering as well as behavioural finance, financial mediation, product innovation, shared values, community building, business strategy and innovation, exponential tech and creation of social capital.

Sustainable Finance and SDG Economics series helps to understand keynotes on international guidelines, guiding accounting and accountability principles, prototyping new developments in triple bottom line investing, cost benefit analysis, integrated financial first plus impact first concepts and impact measurement. Going beyond adjacent fields (like accounting, marketing, strategy, risk management) it integrates the concept of psychology, innovation, exponential tech, choice architecture, alternative economics, blue economy shared values, professions of the future, leadership, human and community development, team culture, impact, quantitative and qualitative measurement, Harvard Negotiation, mediation and complementary currency design using exponential tech and ledger technology. Books in the series contain latest findings from research, concepts for implementation, as well as best practices and case studies for the finance industry.

This book series is indexed in SCOPUS.

Hajer Jarrar · Virginie Hachard · Charbel Salloum ·
Matteo Rossi · Pascal Nguyen
Editors

Sovereign Sustainable Finance

Greening the Treasury-Bonds, Budgets, and Climate Risk

Springer

Editors
Hajer Jarrar
CERIIM—Research Center in Managerial Intelligence and Innovation
Excelia Business School
La Rochelle, France

Charbel Salloum
Métis Lab
EM Normandie Business School
Clichy, France

Pascal Nguyen
Finance
University of Montpellier
Montpellier, France

Virginie Hachard
Métis Lab
EM Normandie Business School
Le Havre, France

Matteo Rossi
Department of Law Economics
Management and Quantitative Methods
University of Sannio
Benevento, Benevento, Italy

ISSN 2522-8285 ISSN 2522-8293 (electronic)
Sustainable Finance
ISBN 978-3-032-33238-7 ISBN 978-3-032-33239-4 (eBook)
https://doi.org/10.1007/978-3-032-33239-4

This Springer imprint is published by the registered company Springer Nature Switzerland AG
The registered company address is: Gewerbestrasse 11, 6330 Cham, Switzerland

Contents

Climate Risk and the Sovereign Balance Sheet: Integrating Exposure, Contingent Liabilities, and Debt Strategy 1
Jean-Michel Sahut

From Nations to Firms: How Sovereign ESG Shapes Corporate Financing Decisions 15
Nouha Ben Arfa, Kaouther Chebbi, and Seyed Alireza Athari

Green and Sustainability-Linked Bonds for Sovereigns and Cities 25
Ekaterina Pirozhkova

Budget Tagging and Climate-Informed Public Financial Management 41
Charbel Salloum

Disaster and State-Contingent Clauses in Sovereign Debt: Trigger Design, Pricing, and Documentation 49
Houanti L'Hocine

Liability Management, Swaps, and Guarantees Climate Risk and Sovereign Green Bonds 61
Miriam R. Aziz

Sub-Sovereign Issuance: PPPs, Pipelines, and Credit Enhancement 71
Charbel Karam

HR and Governance in Debt Offices: Capabilities and Operating Models 83
Enis Mulolli and Xhavit Islami

Risk and Data: Physical/Transition Risk in Fiscal Scenarios and MTFF/MTEF (Medium-Term Expenditure Framework/ Medium-Term Fiscal Framework) 99
Roberta Bajrami, Kaltrina Bajraktari, Besime Ziberi, and Xhavit Islami

Oversight and Assurance in Sovereign Sustainable Finance: SAIs, Internal Audit, and Performance Budgeting 115
Majdi Hassen

Contributors

Seyed Alireza Athari Business School, Holy Spirit University of Kaslik, Jounieh, Lebanon

Miriam R. Aziz Department of Management, INSEEC Grande Ecole, Paris, France

Kaltrina Bajraktari Faculty of Economics, AAB College, Pristina, Kosovo

Roberta Bajrami Faculty of Economics, AAB College, Pristina, Kosovo

Nouha Ben Arfa Excelia Business School, Finance Department, CERIIM, La Rochelle, France

Kaouther Chebbi Finance Department, School of Business, King Faisal University, Al-Ahsa, Saudi Arabia

Majdi Hassen ESSEC Tunis, University of Tunis, Tunis, Tunisia

Xhavit Islami Faculty of Economics, AAB College, Pristina, Kosovo

Charbel Karam Le Cnam Liban, Byblos, Lebanon

Houanti L'Hocine Métis Lab, EM Normandie Business School, Clichy, France

Enis Mulolli Faculty of Economics, AAB College, Pristina, Kosovo

Ekaterina Pirozhkova Métis Lab, EM Normandie Business School, Caen, France;
South African Reserve Bank, Pretoria, South Africa

Jean-Michel Sahut IDRAC Business School, Lyon, France

Charbel Salloum Métis Lab, EM Normandie Business School, Clichy, France

Besime Ziberi Faculty of Economics, AAB College, Pristina, Kosovo

Climate Risk and the Sovereign Balance Sheet: Integrating Exposure, Contingent Liabilities, and Debt Strategy

Jean-Michel Sahut

1 Introduction

Climate risk is no longer a side conversation in sovereign finance. Floods, droughts, heatwaves, storms, and transition policies now affect tax receipts, public investment needs, emergency spending, and the valuation of public assets with enough force to alter debt trajectories. What used to be framed as a long-horizon environmental issue is increasingly handled by treasuries, debt offices, and fiscal risk units because it changes financing needs here and now (Seghini, 2026; Zenios, 2024).

That shift matters because sovereigns do not experience climate risk in only one place. It shows up in damaged infrastructure, weaker growth, higher insurance gaps, support to state-owned enterprises, guarantees to utilities, and pressure on social spending. A balance sheet view makes these channels visible and forces governments to ask a harder question than whether to issue a green bond. The harder question is how climate exposure should reshape debt strategy, annual borrowing plans, investor relations, and the way public institutions coordinate information and accountability (Gonguet et al., 2021; World Bank & International Monetary Fund, 2019).

A balance-sheet perspective matters because sovereign risk is transmitted through several layers at once. There is the obvious layer of damaged public assets and emergency expenditure, but there is also a quieter fiscal layer that sits inside tax capacity, utility support, food and energy subsidies, and public investment deferral. When these channels are mapped together, climate risk stops looking like an environmental side issue and starts looking like a recurring source of fiscal volatility. That broader framing is consistent with recent work on climate and sovereign debt, which shows that exposure, readiness, and institutional execution capacity jointly shape financing conditions rather than any single climate indicator taken in isolation (Arena et al., 2026; Seghini, 2026; Zenios, 2022).

J.-M. Sahut (✉)
IDRAC Business School, Lyon, France
e-mail: jeanmichel.sahut@idrac-bs.fr

H. Jarrar et al. (eds.), *Sovereign Sustainable Finance*, Sustainable Finance,
https://doi.org/10.1007/978-3-032-33239-4_1

This is why thematic borrowing alone cannot solve the problem. A sovereign can issue a green or climate bond and still leave major balance-sheet exposures unmanaged if contingent liabilities, cash buffers, or fiscal reporting remain weak. Chavarría-Flores and Warren (2024) and the World Bank survey of sovereign thematic bonds both make the point indirectly: labeled issuance is most credible when it is embedded in a broader fiscal and debt strategy. Green public financial management reforms push in the same direction because they connect climate commitments to budget preparation, execution, and reporting rather than leaving them at the level of narrative ambition (Battersby et al., 2022; Chavarría-Flores & Warren, 2024; World Bank, 2022).

The chapter develops a conceptual and operational synthesis of recent work on climate risk, sovereign debt, green public financial management, thematic borrowing, and debt-management practice. Its purpose is not to estimate a new climate-debt model, but to show where climate exposure should enter the sovereign decision chain and how ministries of finance can translate that exposure into funding strategy, reporting discipline, and market credibility.

2 Mapping Climate Risk onto the Sovereign Balance Sheet

A sovereign balance sheet perspective begins with assets, not liabilities. Roads, ports, electricity grids, irrigation systems, hospitals, schools, and coastal defenses are public assets whose service value can fall abruptly after climate shocks. Even when physical destruction is limited, recurrent heat stress, water scarcity, and coastal erosion shorten asset lives and raise maintenance costs. The economic loss is therefore broader than headline reconstruction spending. It includes reduced productivity of infrastructure, weaker service delivery, and lower returns on past public investment (Seghini, 2026; Zenios, 2022).

The same logic applies to less visible assets. Tax capacity is partly an asset because it reflects the state's ability to mobilize future revenue. So is the financial health of state-owned enterprises in energy, transport, or water. Climate transition policies can erode the value of carbon-intensive assets or public revenue streams tied to fossil fuel extraction and consumption. For commodity exporters, this creates a balance sheet problem that is easy to understate in traditional debt metrics. The question is not only whether debt service can be paid, but whether the state's asset side is being quietly weakened faster than the liability side is being adapted (Romagnoli & Santini, 2025; Seghini, 2026).

Contingent liabilities make the picture sharper. Governments routinely stand behind public utilities, sub-sovereign entities, infrastructure concessions, development banks, and emergency recovery programs. Those obligations are often absent from the debt stock until a shock materializes. Climate events are precisely the kind of trigger that can turn implicit promises into explicit claims. A drought can force support to public power companies. A cyclone can activate reconstruction transfers to local governments. A failed PPP can push traffic, tariff, or exchange-rate risk back onto the sovereign. The balance sheet view matters because it captures this migration of risk from the periphery to the center (World Bank, 2021).

Recent empirical work confirms that markets are not blind to these dynamics. Using a panel of 31 OECD economies from 2001 to 2020, Arena et al. (2026) show that investors reward stronger environmental management and institutional capacity with lower sovereign borrowing costs. Yet the same study also suggests that large social adaptation requirements can be interpreted as future fiscal rigidity rather than as an immediate resilience asset. That nuance is important. Markets do not simply reward "more climate policy." They reward climate exposure management when it appears fiscally credible and institutionally executable (Arena et al., 2026) (Table 1).

A practical mapping exercise should distinguish at least three types of exposure.

Table 1 Climate risk channels and sovereign balance sheet implications

Climate channel	Where it appears on the balance sheet	Typical debt management consequence	Illustrative example
Physical damage	Public infrastructure and service assets	Higher refinancing needs and reconstruction borrowing	Storm damage to transport and coastal assets in small island states
Transition risk	Revenue base, SOEs, energy systems	Reassessment of medium-term cost-risk targets and funding mix	Carbon-intensive revenue exposure in fuel-importing and exporting economies
Contingent liabilities	Guarantees, PPPs, local entities, utilities	Debt stock can rise suddenly after shocks	Government support to utilities after climate-related service disruptions
Growth and fiscal channel	Tax capacity and primary balance	Higher debt-to-GDP ratios and weaker market access	Lower output and wider deficits after repeated climate events
Financial-sector spillover	Banking system, public financial institutions, guarantees	Pressure for liquidity support or recapitalization during climate stress	Support to publicly exposed lenders after climate-related asset deterioration
Liquidity and cash-buffer stress	Cash position, near-term refinancing profile	Tighter short-term borrowing conditions after shocks	Emergency cash pressure when disaster spending rises before new financing arrives
Monitoring and review	Debt reports, scenario updates, investor dialogue	Adjustment of assumptions before risks compound	Regular update of exposure maps and borrowing assumptions as new evidence emerges

Source Author's own elaboration based on Arena et al. (2026), Seghini (2026), Zenios (2024), and World Bank (2021)

The first is direct sovereign exposure. This includes public assets owned or financed by the state: roads, ports, schools, hospitals, water systems, electricity networks, coastal defences, irrigation systems, and public buildings. These exposures are often visible after disasters, but they should be identified before shocks occur.

The second is revenue exposure. Climate shocks and transition policies can weaken tax bases, reduce export earnings, disrupt tourism, lower agricultural output, or erode revenues from fossil-fuel production and consumption. This matters because tax capacity is an implicit sovereign asset. If the revenue base weakens, debt sustainability can deteriorate even without immediate reconstruction costs.

The third is quasi-fiscal and contingent exposure. This includes liabilities that may sit outside central government debt but can migrate to the budget under stress: state-owned enterprises, utilities, local governments, public-private partnerships, guaranteed loans, development banks, and disaster recovery promises. These exposures are particularly important in countries where public enterprises provide electricity, water, transport, ports, or insurance.

Once these categories are identified, the next step is to connect exposure to response. Table 2 translates the balance-sheet diagnosis into debt-management action.

This table makes the chapter's central point operational. Climate risk should not be treated as a general risk narrative. It should be translated into specific fiscal channels, specific debt-management responses, and specific reporting controls. A government that faces frequent disaster liquidity shocks may need contingent credit lines, disaster clauses, or stronger cash buffers. A government with a well-identified green investment pipeline may benefit from use-of-proceeds borrowing. A sovereign with measurable transition commitments may consider sustainability-linked borrowing. A government with weak budget classification should strengthen budget tagging before issuing labelled debt.

This sequencing is important. Sovereign sustainable finance is most credible when the instrument follows the exposure diagnosis. It is weakest when the instrument is chosen first and the climate logic is assembled afterwards.

3 Bringing Exposure into Debt Strategy and Borrowing Plans

Once climate exposure is mapped, the practical task is to connect it to the debt management cycle. The medium-term debt management strategy is the obvious entry point because it already evaluates the cost-risk consequences of different portfolio compositions. The discipline does not need a parallel "green strategy." It needs climate-informed scenarios within the existing strategy framework. That means testing how alternative shock paths affect financing needs, exchange-rate

Table 2 Climate risk channels and sovereign balance sheet implications

Climate exposure identified	Main fiscal transmission	Debt-management response	Reporting and control implication
Direct public asset damage	Reconstruction spending and service disruption	Resilience borrowing, pre-financing, cash buffers, contingent credit lines	Asset-exposure map and post-disaster spending report
Revenue loss after climate shocks	Weaker primary balance and higher financing needs	More conservative financing assumptions and stress testing	Fiscal-risk statement and macro-fiscal scenario disclosure
SOE or utility exposure	Subsidies, guarantees, recapitalisation, arrears	Contingent-liability monitoring and guarantee limits	SOE risk register and guarantee disclosure
Disaster liquidity shock	Immediate cash pressure before external funds arrive	Disaster clause, pause clause, emergency liquidity buffer	Cash-flow reporting and trigger documentation
Transition risk	Declining fossil-related revenues or higher investment needs	MTDS scenario analysis and funding diversification	Transition-revenue disclosure and investment pipeline tracking
Green investment pipeline	Climate-related capital expenditure needs	Green or sustainability bond	Allocation and impact report
Sovereign-level performance target	Policy-linked climate transition	Sustainability-linked bond or loan	KPI baseline, SPT, external verification
Weak budget classification	Inability to prove eligible spending	Budget tagging before labelled issuance	Audit trail between budget line and financing framework
High refinancing risk after disasters	Market access deterioration	Longer maturities, liability management, guarantees	Debt portfolio stress testing
Sub-sovereign climate exposure	Local reconstruction and service liabilities	On-lending rules, guarantee ceilings, municipal reporting	Local-government fiscal-risk disclosure

Source Author's own elaboration

exposure, interest-rate sensitivity, and the feasible share of domestic versus external debt (World Bank & International Monetary Fund, 2019).

Annual borrowing plans are where this becomes operational. A balance sheet under climate pressure requires tighter forecasting of gross financing needs, more realistic assumptions about disbursements and cash balances, and a clearer view of seasonal or event-related liquidity stress. In countries with concentrated disaster seasons or volatile energy import bills, the calendar of issuance and cash management deserves as much attention as the instrument label. The 2025 IMF and World Bank note on preparing annual borrowing plans makes this point indirectly but clearly: borrowing plans are only credible when they are anchored in detailed cash-flow information, funding constraints, and realistic contingencies (Dissou et al., 2025).

Climate-sensitive public financial management adds another layer. Green PFM is not a separate bureaucracy. At its best, it is an organizing logic that helps ministries connect climate priorities to budgeting, appraisal, reporting, and risk oversight. For debt offices, this matters because investment pipelines, budget tagging, and contingent liability registers determine whether a thematic issuance or resilience-oriented liability operation rests on usable data or on weak promises. The earlier IMF note by Gonguet et al. (2021) and the later operational guidance by Battersby et al. (2022) both stress that climate goals only become fiscally meaningful when they are routed through ordinary public finance systems.

The most successful institutional pattern is therefore coordinated specialization. Debt offices should not own climate policy, and line ministries should not manage sovereign portfolio risk. But the finance ministry needs a mechanism that links climate exposure assessment, fiscal risk analysis, project selection, and debt execution. Without that bridge, the sovereign may sound climate-aware to investors while still borrowing against incomplete information. In practice, mature arrangements give the debt office early access to project pipelines, contingent liability updates, disaster financing protocols, and macro-fiscal stress scenarios rather than involving it only at the end of a bond transaction (Battersby et al., 2022; World Bank & International Monetary Fund, 2014).

Once exposure is identified, the discipline has to move into the routine debt cycle. The medium-term debt management strategy is the natural anchor because it already forces governments to compare cost and risk across different portfolio choices. Climate information belongs inside that framework rather than beside it. Debt managers need scenarios for disaster shocks, revenue shortfalls, exchange-rate pressure, and higher investment needs, and those scenarios need to be reflected in annual borrowing plans and cash management assumptions. IMF guidance on annual borrowing plans and green PFM guidance both point toward the same operational lesson: climate considerations become useful when they alter baseline financing decisions, not when they sit in a parallel document (Battersby et al., 2022; Dissou et al., 2025; World Bank and International Monetary Fund, 2019).

Investor communication becomes part of that architecture as well. Markets tend to price coherence, repeat issuance discipline, and reporting quality more reliably than they price broad sustainability rhetoric. A sovereign that can explain how climate exposure affects its borrowing strategy, contingency planning, and use-of-proceeds reporting is easier to assess than one that treats climate finance as a purely political narrative. This is why sovereign investor relations matter even for balance-sheet management. Regular dialogue, transparent documentation, and clear disclosure of methodology reduce uncertainty about what the government is actually trying to manage and how it will report outcomes over time (Knight & Northfield, 2020; Romagnoli & Santini, 2025; World Bank, 2022).

Data limitations make conservative integration especially important. Climate exposure is often easier to discuss than to quantify consistently, and governments rarely have complete information on local assets, service networks, or contingent claims. That does not justify postponing integration. It suggests a sequencing rule instead. Debt offices should start with the exposures they can already identify well enough to affect financing decisions, then improve granularity over time. Waiting for perfect data would leave borrowing strategies anchored to an outdated view of risk for too long (Battersby et al., 2022; World Bank and International Monetary Fund, 2019).

4 What Markets Are Already Telling Sovereign Issuers

Sovereign thematic bonds offer a useful reading of market preferences, provided they are not romanticized. Recent evidence from emerging economies suggests that sovereign green bonds can improve market visibility, diversify the investor base, and strengthen climate-finance signaling when they are embedded in a broader public finance framework. Owusu-Mante et al. (2025) show that Chile, Egypt, Indonesia, and Nigeria all met targeted issue amounts, with oversubscriptions ranging from negligible levels in early Nigerian experience to exceptionally strong demand in Chile. In Chile's first 2019 sovereign green bond, demand reached roughly 12.8 times the amount offered. Egypt's debut green issuance attracted an order book around seven times the initial offer, allowing both an upsized issue and a lower coupon. Those are not symbolic outcomes. They affect cost, tenor, and investor composition (Owusu-Mante et al., 2025).

Indonesia's green sukuk program is especially instructive because it shows continuity rather than one-off experimentation. By year-end 2022, Indonesia had raised about US$6.9 billion through repeated sovereign green issuances, with multiple offerings oversubscribed and supported by regular allocation and impact reporting (Ministry of Finance, Republic of Indonesia, 2023; Owusu-Mante et al., 2025). Chile provides a complementary lesson from the other side of the market. Its reporting architecture, repeated issuance, and broad investor outreach helped position labeled borrowing as part of debt management practice rather than as a communications exercise (Chile Ministry of Finance, 2019).

Still, a green label does not neutralize sovereign climate risk by itself. Nevatia (2026) argues that sovereign green bonds can function as risk-mitigating instruments in emerging markets, yet the benefits depend on broader credibility conditions. The same theme appears in Dell'Atti et al. (2022), who find that sovereign green bond issuance interacts with perceived country value and risk rather than bypassing them. This is why the balance sheet lens remains useful. A sovereign that issues a green bond while leaving contingent liabilities, weak reporting systems, or disaster liquidity gaps unaddressed may gain a pricing benefit at the margin but still retain a fragile funding structure (Dell'Atti et al., 2022; Nevatia, 2026).

Markets also appear to reward clarity on use of proceeds, verification, and institutional follow-through. World Bank work on sovereign green, social, and sustainability bonds emphasizes that emerging market issuers are more likely to gain durable market benefits when issuance frameworks are linked to budget systems, reporting, and credible governance. Chavarría-Flores and Warren (2024) add an important political point: sovereign climate bonds in developing countries work best when they support a just transition strategy that is fiscally legible rather than rhetorically expansive. Investors can absorb complexity. What they resist is ambiguity about how climate goals connect to debt service capacity and policy execution (Chavarría-Flores & Warren, 2024; World Bank, 2022).

The market evidence on sovereign green bonds reinforces that caution. A green label can support credibility and sometimes reduce perceived risk, but pricing effects are not uniform. Nevatia (2026) finds meaningful heterogeneity between advanced and emerging markets, with evidence that green labeling in riskier emerging settings can lower informational asymmetry and support pricing. Even so, that benefit is not automatic. Earlier evidence from European issuers and more recent work on emerging-economy practice both suggest that the quality of project selection, reporting, and institutional follow-through still matters. In short, green issuance can help, but only where the sovereign can sustain the information and governance burden that comes with it (Dell'Atti et al., 2022; Nevatia, 2026; Owusu-Mante et al., 2025).

Repeated issuance gives a useful signal here. Chile and Indonesia are important not because they prove that all sovereign green bonds lower financing costs, but because they show how repeated reporting can gradually turn climate borrowing into an operating routine rather than a one-off innovation. Allocation and impact reporting in Chile and the green sukuk reporting cycle in Indonesia both illustrate how disclosure practices can mature over time. That experience matters for balance-sheet management because it builds internal habits around eligibility, data collection, and post-issuance monitoring. The process is operationally demanding, yet it is far closer to what ministries of finance actually need than symbolic green financing with weak reporting follow-through (Chile Ministry of Finance, 2019; Ministry of Finance, Republic of Indonesia, 2023; Nevatia, 2026).

Markets also reward repetition. A one-off framework can attract attention, but repeated disclosure, comparable methodologies, and stable governance arrangements are what gradually turn climate-related claims into something investors can price with greater confidence. That is why reporting practice matters almost as much as the initial transaction. When issuers publish allocation and impact reports on time and can explain how those reports connect to broader fiscal management, they reduce the perception that sustainable finance is separate from debt discipline. In practical terms, that reduction in uncertainty is part of the sovereign's financial resilience (Knight & Northfield, 2020; World Bank, 2022).

5 Choosing the Right Sovereign Sustainable Finance Instrument

Instrument choice should follow the diagnosis of exposure, data capacity, and policy credibility. Different sovereign sustainable finance instruments solve different problems. A green bond is not a disaster-liquidity instrument. A sustainability-linked bond is not a substitute for budget tagging. A guarantee does not replace fiscal-risk disclosure. A debt-for-climate swap does not eliminate the need for public financial management controls.

Table 3 summarises the main instruments and the conditions under which they are most appropriate.

This table highlights a sequencing discipline. If the main problem is weak identification of climate expenditure, the priority is budget tagging and reporting, not a complex labelled instrument. If the main risk is a post-disaster liquidity shock, the relevant tools may be disaster clauses, pause clauses, cash buffers, or contingent credit lines. If the sovereign has a credible investment pipeline, use-of-proceeds borrowing may be appropriate. If the sovereign wants to commit to measurable policy outcomes, a sustainability-linked instrument may be considered, but only where KPIs are material and targets credible.

A sustainability-linked sovereign instrument requires particular caution. Unlike use-of-proceeds bonds, proceeds are often used for general budgetary purposes. Credibility therefore depends on the quality of the KPI architecture. The KPI must be material to the sovereign's climate or resilience strategy. The baseline must be transparent. The sustainability performance target must be ambitious relative to historical performance and policy commitments. The verification process must be independent. The financial consequence must be meaningful enough to signal commitment without creating perverse incentives.

A green or sustainability bond requires a different discipline. The key issue is whether the eligible expenditure can be identified, tracked, reported, and audited. Budget tagging therefore becomes central. If a government cannot link bond proceeds to eligible budget lines, ministries, programmes, or projects, the labelled instrument rests on weak foundations. The credibility of the issuance depends on the traceability of the expenditure and the reliability of post-issuance reporting.

Disaster and pause clauses serve another function. They are not primarily climate-investment instruments. They are state-contingent debt features that protect liquidity after shocks. Their value depends on whether the trigger is objective, the deferral mechanism is clear, and investors understand the pricing implications. These clauses may be particularly useful for countries exposed to severe natural disasters, but they must be integrated into the broader debt strategy rather than treated as isolated innovations.

Guarantees and swaps also belong to the sovereign sustainable finance toolkit. Guarantees can reduce borrowing costs, extend maturities, or mobilise private capital, but they also create risk-sharing and moral-hazard questions. Debt-for-climate or debt-for-nature swaps can create fiscal space and support conservation or resilience spending, but only if governance is strong, additionality is credible, and monitoring is reliable.

The practical rule is simple: the instrument should match the balance-sheet problem. Sustainable finance becomes credible when instrument design follows exposure diagnosis, not when the label precedes the analysis.

Table 3 Matching sovereign sustainable finance instruments to policy and balance-sheet needs

Instrument	Best suited when	Main credibility condition	Main risk if poorly designed
Green bond	Eligible green expenditure is clearly identifiable in the budget	Robust use-of-proceeds framework, allocation reporting, impact reporting	Broad expenditure categories and weak reporting
Sustainability bond	Climate and social expenditures are jointly relevant	Clear classification of environmental and social use of proceeds	Overly broad labelling
Sustainability-linked bond	Sovereign-level performance can be measured credibly	Material KPIs, robust baselines, ambitious targets, meaningful financial consequences	Weak KPIs or non-ambitious targets
Green sukuk	Eligible assets and expenditures can be structured consistently with Islamic finance principles	Asset linkage and credible allocation reporting	Formal compliance without strong impact evidence
Disaster clause	The sovereign faces material disaster-related liquidity risk	Objective trigger, clear payment-deferral terms, investor understanding	Ambiguous trigger or weak pricing logic
Pause clause	Temporary debt-service relief is valuable after shocks	Transparent activation conditions and contractual clarity	Perceived weakening of repayment discipline
Contingent credit line	Rapid liquidity is needed after disasters	Pre-agreed access, eligibility conditions, fiscal discipline	Reliance on facility without broader resilience planning
Guarantee	The sovereign needs to lower cost or extend maturity	Clear guarantor role, risk-sharing structure, reporting conditions	Hidden fiscal cost or moral hazard
Debt-for-climate or debt-for-nature swap	Debt relief can be linked to credible climate or conservation spending	Strong governance, conservation fund, monitoring and audit	Complex transaction with weak fiscal additionality
Budget tagging	The state needs to prove climate-related expenditure	Clear methodology, budget-line traceability, auditability	Over-classification or inconsistent tagging
Blended finance	High-impact projects lack standalone commercial viability	Clear public-private risk allocation and outcome metrics	Subsidising weak projects without accountability

Source Author's own elaboration

6 A Practical Operating Agenda for Ministries of Finance and Debt Offices

A ministry of finance seeking to integrate climate risk into sovereign debt management should begin with a practical operating agenda rather than a complex institutional redesign. The first priority is to build a usable exposure map. This map should identify direct public assets, revenue exposures, contingent liabilities, SOEs, utilities, local governments, PPPs, disaster-prone sectors, and major climate-sensitive investment needs. The purpose is not to produce a perfect model. It is to identify exposures that are material enough to affect financing decisions.

The second priority is to translate exposure into fiscal-risk and debt-strategy tools. Climate scenarios should appear in medium-term debt strategies, fiscal-risk statements, annual borrowing plans, and cash-management assumptions. Governments do not need to forecast every climate outcome. They need to test plausible stress cases and identify how those cases would affect gross financing needs, liquidity, maturity structure, exchange-rate exposure, and market access.

The third priority is to align instrument choice with institutional readiness. If budget data are weak, the government should strengthen budget tagging and reporting before issuing a labelled bond. If disaster liquidity is the main concern, the focus should be cash buffers, contingent financing, or disaster clauses. If a credible green investment pipeline exists, a use-of-proceeds instrument may be appropriate. If a sovereign-level transition target can be measured and verified, a sustainability-linked structure may be considered.

The fourth priority is to integrate investor communication into debt management. Sovereign investor relations should explain not only the labelled instrument but also the broader climate-risk management strategy. Investors need to understand how climate exposure is identified, how it enters borrowing plans, how proceeds or KPIs are tracked, and how the government will report outcomes.

The fifth priority is review. Climate evidence, market expectations, and fiscal exposure change over time. Ministries of finance should update climate-related debt assumptions regularly, review eligible expenditure categories, reassess contingent liabilities, and improve data coverage. The balance-sheet map should be a living tool rather than a one-off diagnostic.

In practice, the operating agenda can be summarised as follows:

1. identify climate-related assets, revenues, and contingent liabilities;
2. classify exposures by fiscal transmission channel;
3. incorporate climate scenarios into MTDS and annual borrowing plans;
4. strengthen budget tagging and expenditure traceability;
5. select instruments according to exposure and data capacity;
6. define KPIs, targets, MRV, and verification arrangements;
7. prepare audit-ready controls;
8. communicate the strategy to investors;
9. review assumptions and reporting annually.

This agenda does not require debt offices to become climate ministries. It requires existing fiscal and debt-management tools to reflect climate-linked financing pressures more realistically.

7 Conclusion

Climate risk matters for sovereign debt because it reshapes the balance sheet that supports repayment capacity. Physical shocks destroy or impair public assets, weaken tax revenues, increase emergency expenditure, and create liquidity pressure. Transition risks can erode parts of the fiscal base, alter the value of state-owned enterprises, raise public investment needs, and affect borrowing conditions. Contingent liabilities may also become explicit with little warning when climate shocks affect utilities, sub-sovereign entities, public-private partnerships, public banks, or disaster recovery programmes. When these channels are ignored, debt strategies may appear more resilient on paper than they are in practice.

This chapter has argued that sovereign sustainable finance should begin with exposure mapping rather than labelled issuance alone. Green bonds, sustainability bonds, sustainability-linked bonds, green sukuk, disaster clauses, pause clauses, guarantees, contingent credit lines, and debt-for-climate or debt-for-nature swaps can all be useful instruments. However, they are credible only when they respond to a clearly identified sovereign balance-sheet problem. The instrument should follow the diagnosis, not replace it. A green bond is appropriate when eligible expenditure is identifiable and reportable. A sustainability-linked bond is defensible when a sovereign-level target is material, measurable, ambitious, and externally verifiable. Disaster or pause clauses add value when the main risk is post-shock liquidity pressure. Guarantees or swaps can help rebalance sovereign or municipal balance sheets, but only when fiscal risks, governance arrangements, and monitoring responsibilities are transparent.

The balance-sheet perspective also clarifies why thematic borrowing alone cannot solve the climate-debt problem. A sovereign can issue a green or climate bond and still leave major exposures unmanaged if contingent liabilities, disaster liquidity needs, budget tagging, cash buffers, or fiscal-risk reporting remain weak. Conversely, a sovereign that integrates climate exposure into medium-term debt strategy, annual borrowing plans, fiscal-risk statements, green public financial management, investor communication, and audit-ready controls can strengthen credibility even before using more complex instruments. The key issue is therefore not whether a transaction carries a sustainable label, but whether climate risk is visible in the ordinary machinery of public finance.

For ministries of finance and debt-management offices, the practical implication is direct. Climate risk should enter the medium-term debt management strategy through scenarios that test disasters, transition shocks, revenue losses, refinancing pressure, exchange-rate movements, and contingent liabilities. It should enter annual borrowing plans through realistic assumptions about gross financing needs, cash balances, disbursement timing, and emergency liquidity. It should enter fiscal-risk reporting through clearer disclosure of state-owned enterprise exposure, public guarantees, local-government vulnerabilities, and disaster-related commitments. It should also enter budget systems through climate tagging, expenditure traceability, project appraisal, allocation reporting, and audit trails.

For sovereign issuers, credibility depends on moving from climate narrative to verifiable financing discipline. If the chosen instrument is a use-of-proceeds green or sustainability bond, the issuer must identify eligible budget lines, define selection criteria, manage proceeds, report allocation, and provide credible impact indicators. If the instrument is sustainability-linked, the issuer must show that the KPI is material to the country's transition or resilience pathway, that the baseline is transparent, that the target is ambitious, and that verification is independent. If the instrument contains a disaster clause or pause clause, the trigger, deferral mechanism, contractual implications, and investor treatment must be clear. In each case, the label is only as strong as the institutional system that supports it.

A balance-sheet view also helps explain why climate risk differentiates sovereigns across regions and institutional settings. In countries with stronger environmental management, adaptation capacity, and fiscal institutions, climate exposure may remain serious but is less likely to translate immediately into a market penalty. In more fragile settings, the same type of shock can tighten financing conditions quickly, reduce investment capacity, and weaken future resilience. This feedback loop is central to the climate-sovereign debt problem. Climate risk is not only a direct source of physical or fiscal loss. It can also become a financing constraint that reduces the state's ability to respond to future shocks. This is why debt strategy and climate strategy can no longer be written as separate conversations (Arena et al., 2026; Zenios, 2024).

The broader contribution of this chapter is to reposition sovereign sustainable finance as a discipline of balance-sheet management. The market will mature not because more sovereigns issue labelled instruments, but because more governments can demonstrate how climate exposure is identified, financed, managed, reported, verified, and reviewed. A mature sovereign response must finance climate priorities where appropriate while protecting the balance sheet from climate-related surprises. That requires a living exposure map, climate-sensitive fiscal assumptions, updated contingent-liability registers, realistic borrowing plans, clear instrument selection, credible KPIs, audit-ready reporting systems, and regular review.

Sovereign sustainable finance becomes credible when it is evidence-based, operationally grounded, institutionally coordinated, and repeatedly tested against one central question: does climate risk visibly shape the way the state borrows, reports, manages liquidity, and protects its future repayment capacity?

References

Arena, C., Basiglio, S., Comandé, A., Gozgor, G., & Li, J. (2026). The impact of climate risk on sovereign debt: Evidence from OECD economies. *Journal of Environmental Management, 402*, 128851. https://doi.org/10.1016/j.jenvman.2026.128851

Battersby, B., Gonguet, F., Wendling, C. P., Charaoui, J., & Petrie, M. (2022). *How to make the management of public finances climate-sensitive—"green PFM"* (IMF How To Notes, 2022(006)). https://doi.org/10.5089/9798400221682.061

Chavarría-Flores, A., & Warren, P. (2024). Sovereign climate bonds: Policy innovation for just transitions in developing countries. *Journal of Climate Finance, 7*, 100041. https://doi.org/10.1016/j.jclimf.2024.100041

Chile Ministry of Finance. (2019). *2019 report: Allocation and environmental impact (green bonds).* Ministry of Finance, Government of Chile.

Dell'Atti, S., Di Tommaso, C., & Pacelli, V. (2022). Sovereign green bond and country value and risk: Evidence from European Union countries. *Journal of International Financial Management & Accounting, 33*(3), 505–521. https://doi.org/10.1111/jifm.12155

Dissou, O., Knight, J., Tamene, M., Secunho, L., & Rivetti, D. (2025). *Preparing an annual borrowing plan* (IMF Technical Notes and Manuals, 2025(012)). https://doi.org/10.5089/9798229011433.005

Gonguet, F., Wendling, C. P., & Battersby, B. (2021). *Climate-sensitive management of public finances—"green PFM"* (IMF Staff Climate Notes, 2021(002)). https://doi.org/10.5089/9781513583044.066

Knight, J., & Northfield, B. (2020). *Sovereign investor relations: From principles to practice* (IMF Working Papers, 2020(204)). https://doi.org/10.5089/9781513556369.001

Ministry of Finance, Republic of Indonesia. (2023). *Green sukuk allocation and impact report 2023.* Ministry of Finance, Republic of Indonesia.

Nevatia, V. (2026). Sovereign green bonds: Risk-mitigating sustainability instruments in emerging markets. *Energy Economics, 155*, 109173. https://doi.org/10.1016/j.eneco.2026.109173

Owusu-Mante, S., Chi, K. H.-K., & Gopalakrishnan, T. (2025). Sovereign green bonds in practice: Effectiveness as a climate finance instrument in emerging economies. *Climate Policy.* Advance online publication. https://doi.org/10.1080/14693062.2025.2586572

Romagnoli, S., & Santini, A. (2025). Climate risk and sovereign debt: Country-level exposures and scarcity effects in green bonds. *Decisions in Economics and Finance.* Advance online publication. https://doi.org/10.1007/s10203-025-00554-5

Seghini, C. (2026). The macroeconomics of climate change and public debt sustainability: Mapping transition and physical risks through output growth, primary balances, and real rates. *Journal of Economic Surveys.* Advance online publication. https://doi.org/10.1111/joes.70060

World Bank. (2021). *Debt management performance assessment methodology: 2021 edition.* World Bank.

World Bank. (2022). *Sovereign green, social and sustainability bonds: Unlocking the potential for emerging markets and developing economies.* World Bank.

World Bank, & International Monetary Fund. (2014). *Revised guidelines for public debt management.* World Bank.

World Bank, & International Monetary Fund. (2019). *Developing a medium-term debt management strategy (MTDS): Updated guidance note for country authorities.* World Bank and International Monetary Fund.

Zenios, S. A. (2022). The risks from climate change to sovereign debt. *Climatic Change, 172*, 30. https://doi.org/10.1007/s10584-022-03373-4

Zenios, S. A. (2024). The climate-sovereign debt doom loop: What does the literature suggest? *Current Opinion in Environmental Sustainability, 67*, 101414. https://doi.org/10.1016/j.cosust.2024.101414

From Nations to Firms: How Sovereign ESG Shapes Corporate Financing Decisions

Nouha Ben Arfa, Kaouther Chebbi, and Seyed Alireza Athari

1 Introduction

Over the past decades, environmental, social, and governance (ESG) considerations have gradually shifted from the periphery to the center of both public policy and corporate strategy (Crifo et al., 2017; Zhang et al., 2022). ESG refers to environmental, social, and governance factors that measure sustainability and institutional quality at both the firm and country levels. Governments are increasingly including sustainability goals within national economic frameworks, thereby influencing regulatory environments, financial systems, and corporate conduct.

Simultaneously, ESG has attracted significant attention from investors, financial institutions, and international organizations. These findings also indicate that integrating ESG into financial decision-making is not just a passing trend but a part of a larger structural shift in financial systems. As sustainability considerations become more deeply embedded in regulatory frameworks and investment strategies, their impact on corporate behavior is expected to increase.

In this context, sovereign ESG could become more important in shaping not only financing conditions but also broader economic results.

In many situations, ESG considerations are now seen as essential parts of financial decision making, not optional.

N. Ben Arfa (✉)
Excelia Business School, Finance Department, CERIIM, La Rochelle, France
e-mail: benarfan@excelia-group.com

K. Chebbi
Finance Department, School of Business, King Faisal University, Al-Ahsa, Saudi Arabia
e-mail: kchebbi@kfu.edu.sa

S. A. Athari
Business School, Holy Spirit University of Kaslik, Jounieh, Lebanon
e-mail: ali_athari@yahoo.com

H. Jarrar et al. (eds.), *Sovereign Sustainable Finance*, Sustainable Finance,
https://doi.org/10.1007/978-3-032-33239-4_2

Despite this growing significance, most current studies mainly concentrate on firm level ESG practices (Ioannou & Serafeim, 2012; Khan et al., 2016). Less attention has been given to sovereign ESG and its effects on corporate financial decisions. This gap is especially meaningful because firms do not operate alone. They develop within institutional environments that influence their access to capital, their risk exposure, and ultimately their financing strategies.

In emerging and resource dependent economies, such as those in the GCC region, institutional frameworks are especially important. Sovereign ESG can therefore influence corporate behavior in ways that can't be fully understood through firm-level analyses alone.

This chapter addresses this gap by examining how sovereign ESG influences firms' capital structure decisions in GCC countries. Instead of offering a purely technical analysis, the chapter takes a synthesis-oriented approach, emphasizing the interpretation of empirical findings and their broader implications. It also contributes to the literature by summarizing how sovereign ESG impacts corporate financing decisions, especially in resource-dependent economies. This perspective is especially relevant given the increasing global focus on sustainability and responsible investment. As ESG considerations become more embedded in international financial markets, the interaction between sovereign-level sustainability and firm-level financial decisions is likely to grow stronger. Understanding these dynamics is therefore crucial not only for academic research but also for policy development and financial practice.

2 Sovereign ESG and Corporate Financing: Key Mechanisms

Sovereign ESG performance indicates the overall quality of a country's environmental policies, social conditions, and governance frameworks (Gratcheva et al., 2021; Zhang et al., 2022). These aspects shape the institutional environment in which firms operate and affect financial decisions through several interconnected channels.

A primary mechanism relates to risk and institutional credibility. Countries with strong governance and stable social environments tend to have lower uncertainty levels (Capelle-Blancard et al., 2016; Hübel & Scholz, 2020). This decrease in sovereign risk usually leads to better financing conditions for firms (Djankov et al., 2008; La Porta et al., 1998). In such cases, lenders see lower default risk, which makes it easier for firms to access debt markets and may lead them to increase leverage.

A second mechanism involves financial development. ESG focused policies, through transparency, accountability, and institutional strengthening, help build investor confidence (Levine, 2005; Ng et al., 2020; Wang et al., 2019). Over time, this process fosters the growth of deeper and more efficient financial markets. Firms operating in such environments are therefore more likely to access external financing under favorable conditions.

A third channel functions through alignment effects. Firms in high ESG settings may modify their strategies to satisfy regulatory demands or investor preferences. This alignment can impact not only their operational choices but also their financing approaches, especially when access to capital depends on ESG related standards.

Additionally, sovereign ESG can affect firms' cost of capital. A stronger ESG profile at the country level might lower risk premiums, boost sovereign credit ratings, and increase investor confidence. This, in turn, can reduce borrowing costs (Goss & Roberts, 2011) and encourage firms to depend more on external financing.

Another vital mechanism involves institutional enforcement. Effective governance systems enhance contract enforcement and regulatory effectiveness, which reduces uncertainty for lenders and investors. This leads to a more predictable financial environment, encouraging long-term financing decisions.

However, ESG should not be viewed as a homogeneous construct. Its different dimensions may have distinct, and sometimes contrasting effects. Environmental policies, for instance, often entail short-term compliance costs. These costs may be particularly significant in capital-intensive or resource-based industries. By contrast, improvements in governance and social conditions tend to have more immediate and positive effects by reducing uncertainty and strengthening institutional trust.

These mechanisms can also interact with each other. For example, improvements in governance can strengthen the effectiveness of environmental and social policies, thereby boosting overall institutional credibility (Capelle-Blancard et al., 2016; Hübel & Scholz, 2020). Likewise, stronger social conditions may support political stability, which then helps create more predictable economic environments.

Moreover, sovereign ESG can indirectly influence firms through the banking sector. In countries with stronger ESG frameworks, financial institutions might adopt more rigorous risk assessment methods and incorporate sustainability criteria into lending decisions. This can impact both the availability and the cost of credit, further shaping corporate financing strategies.

Finally, it is crucial to recognize that the impact of sovereign ESG can change over time. While some effects, such as enhancements in governance, may offer immediate benefits, others especially environmental reforms, might take longer to show results. This dynamic view emphasizes the need to consider both short-term and long-term effects when examining the connection between ESG and corporate finance.

The conceptual link between sovereign ESG performance and corporate leverage is shown in Fig. 1.

3 Empirical Insights from GCC Countries

The empirical analysis uses firm level panel data from GCC countries spanning 2007–2022. The relationship between sovereign ESG and corporate leverage is analyzed using standard panel regression methods, controlling for firm-specific factors such as size, profitability, and asset structure, as well as for time and sector effects. The findings highlight several consistent and economically significant patterns.

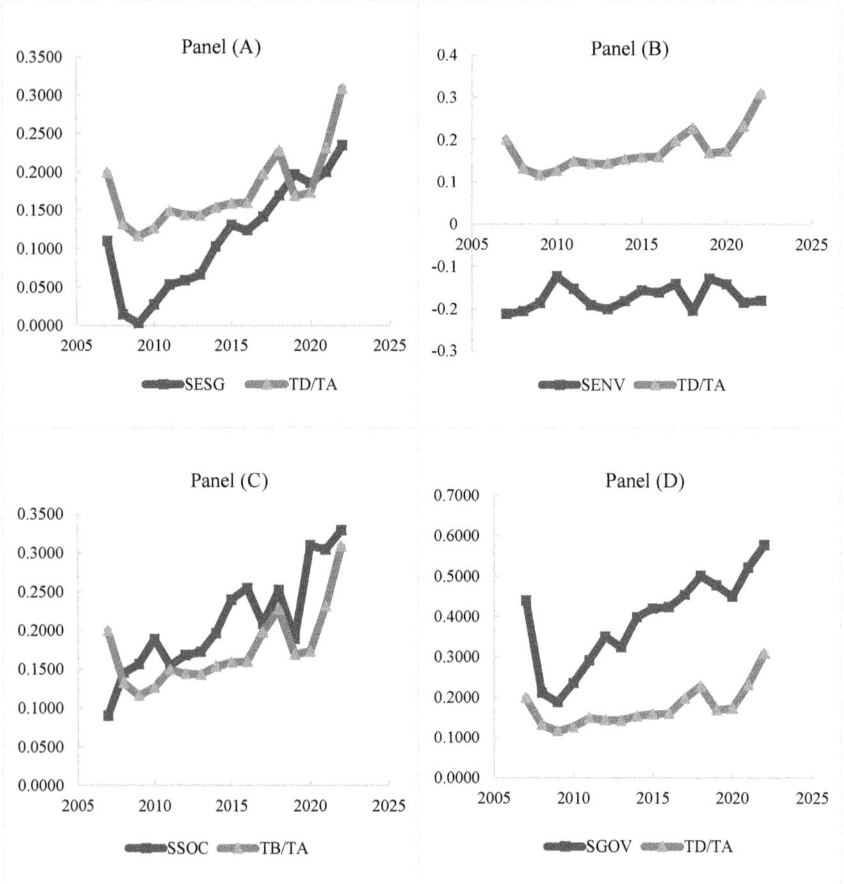

Fig. 1 Sovereign ESG and corporate leverage (conceptual). Panel A: Link between overall sovereign ESG performance and corporate leverage. Panel B: Different impacts of ESG components on leverage. *Note* ESG stands for environmental, social, and governance. SESG represents sovereign ESG; SENV indicates environmental performance; SSOC reflects social performance; SGOV measures governance quality; TD/TA is total debt to total assets; TB/TA is total liabilities to total assets

First, sovereign ESG performance is positively linked to corporate leverage. Firms operating in countries with stronger ESG frameworks tend, on average, to depend more on debt financing. One possible interpretation is that improved institutional environments lower financing frictions and improve access to credit. In this way, ESG can be seen as a signal of macroeconomic stability and institutional credibility.

Second, ESG dimensions show varied effects. Governance and social factors are positively linked to firms' borrowing behavior. These dimensions reflect institutional quality, legal protections, and social stability elements that generally improve firms' creditworthiness and lower lenders' risk perception.

Environmental performance, however, seems to have a more subtle role. Sometimes, it is linked to lower leverage. At first, this might seem surprising given the growing focus on sustainability. But it probably reflects short-term costs of adjusting to environmental rules. Firms may need to make extra investments, meet regulatory standards, or deal with operational limits that temporarily reduce their ability to take on more debt.

This interpretation aligns with the idea that sustainability transitions involve phases of adjustment. While ESG improvements may boost financial conditions over the long term, their short-term effects can vary depending on the dimension evaluated.

The negative association seen with environmental performance might also indicate transitional dynamics. Firms in economies going through environmental reforms may encounter regulatory uncertainty and changing standards, which can affect their financing strategies.

Third, firm-level characteristics remain central determinants of capital structure. Larger firms and those with significant intangible assets are more likely to access debt markets, benefiting from scale effects and reputational advantages. Conversely, firms with higher profitability and stronger internal cash flows tend to rely less on external financing, consistent with established corporate finance theories (Myers, 1984; Rajan & Zingales, 1995).

Taken together, these findings suggest that sovereign ESG complements, rather than replaces, traditional determinants of capital structure. More broadly, they emphasize the importance of considering institutional environments when analyzing corporate financial behavior.

These main findings are summarized in Table 1.

Table 1 Summary of important findings

Factor	Measurement	Effect on leverage	Interpretation
Sovereign ESG	Corporate ESG score at the country level (SESG)	Positive	Strong ESG environments reduce financing constraints
Environmental	Environmental score (SENV)	Negative	Compliance costs may limit borrowing
Social	Social score (SSOC)	Positive	Social stability ilmproves firm credibility
Governance	Governance score (SGOV)	Positive	Institutional quality lowers perceived risk
Firm size	Natural logarithm of total assets	Positive	Larger firms access debt more easily
Profitability	Return on assets (ROA) or operating income over assets	Negative	Internal funds reduce need for borrowing
Covid-19	Dummy variable equal to 1 for pandemic period	Negative	Crisis reduces reliance on ESG-driven borrowing

Note SESG stands for sovereign ESG score; SENV indicates environmental performance; SSOC refers to social performance; SGOV represents governance quality; ROA denotes return on assets

These results indicate that sovereign ESG functions as both a facilitator and a limiter, depending on the dimension evaluated and the economic environment.

These findings also offer insights into the broader influence of institutional environments on corporate behavior. They imply that firms adjust their financial strategies not only based on internal factors but also considering external constraints and opportunities established at the country level.

In particular, the positive link between sovereign ESG and leverage shows that sustainability frameworks can help promote financial development. By reducing uncertainty and increasing transparency, ESG supports a more efficient allocation of capital.

Simultaneously, the varied effects seen across ESG dimensions highlight the complexity of sustainability transitions. While governance and social improvements tend to produce relatively quick benefits, environmental policies might involve adjustment costs that are more significant in the short term. This underscores that ESG should not be viewed as a single dimensional concept but as a collection of interconnected factors with different economic implications.

4 Sectoral and Crisis-Specific Impacts

The impact of sovereign ESG on corporate financing varies by sector, reflecting differences in financing requirements and economic frameworks.

Capital-intensive industries like energy, industrials, and real estate show a stronger reaction to ESG improvements. These sectors depend heavily on external financing and are therefore more affected by changes in borrowing conditions. Enhancements in ESG performance can lower borrowing costs and make it easier to access credit, encouraging firms to boost leverage.

In contrast, sectors with lower capital intensity or greater reliance on internal financing exhibit weaker relationships. This indicates that ESG-related improvements in financial conditions are especially important when firms depend on external capital markets.

The differential impact across sectors also indicates that ESG effects are mediated by industry characteristics. Firms operating in highly regulated or environmentally sensitive industries may respond differently to ESG improvements compared to those in less regulated sectors.

The analysis also highlights the role of crisis periods, particularly the COVID-19 pandemic. During this period, firms faced declining revenues, increased uncertainty, and significant liquidity constraints. Many firms increased their reliance on debt financing to sustain operations.

However, the positive relationship between sovereign ESG and leverage weakened during the crisis. In such times, firms tend to prioritize short-term liquidity needs over the structural benefits linked to ESG factors. This suggests that ESG-related advantages may be less visible during periods of high uncertainty.

This finding does not imply that ESG becomes irrelevant during crises. Rather, it suggests that its effects operate over longer time horizons, contributing to resilience and stability in the medium to long term.

5 Discussion and Implications for Stakeholders

The findings provide several insights that extend beyond the empirical results.

From a theoretical perspective, they suggest that traditional capital structure theories, such as trade-off and pecking-order models can be enriched by incorporating institutional factors. While firm level determinants remain central, they operate within a broader context shaped by sovereign ESG.

These results contribute to the literature by highlighting the role of sovereign ESG as a multidimensional determinant of corporate financial behavior. They suggest that institutional quality, sustainability policies, and governance structures should be considered alongside traditional financial variables (Ioannou & Serafeim, 2012).

From a policy perspective, the results highlight the importance of balanced ESG strategies. Strengthening governance and social frameworks appears to improve financial conditions and facilitate firms' access to capital. Environmental policies are equally important but may require careful implementation to avoid imposing excessive short-term constraints, particularly in sectors facing high compliance costs.

For policymakers, this implies that ESG policies should be designed not only with sustainability objectives in mind but also with consideration for their economic and financial implications.

For investors, sovereign ESG provides valuable information about the institutional environment in which firms operate. Incorporating ESG at the country level can enhance risk assessment and improve portfolio allocation decisions (Friede et al., 2015), particularly in emerging markets.

For creditors, ESG considerations can complement traditional credit risk analysis. Firms operating in stronger ESG environments may benefit from more stable economic conditions, leading to lower perceived risk.

More broadly, ESG can be understood not only as a sustainability concern but also as a key factor shaping financial conditions, capital allocation, and corporate behavior.

These findings also suggest that the role of ESG extends beyond sustainability considerations and should be integrated into mainstream financial analysis.

They highlight the need for a more holistic approach to corporate finance, where institutional and sustainability factors are jointly considered.

This perspective is particularly relevant in emerging economies, where institutional environments play a central role in shaping financial outcomes.

These implications also highlight the importance of coordination between different stakeholders. Policymakers, investors, and financial institutions do not operate independently, and their decisions collectively shape financial markets and corporate behavior. A coordinated approach to ESG integration can therefore enhance its effectiveness and reduce potential trade-offs.

In addition, the findings suggest that ESG considerations may become increasingly important in the context of global financial integration. As international investors pay greater attention to sustainability indicators, countries with stronger ESG performance may benefit from improved access to capital at both sovereign and corporate levels.

Finally, these results emphasize the need for continuous improvement in ESG measurement and reporting. Reliable and comparable ESG indicators are essential to ensure that sustainability considerations are effectively integrated into financial decision-making processes.

6 Conclusion

This chapter shows that sovereign ESG plays a meaningful role in shaping corporate financing decisions. By influencing institutional quality, financial market development, and risk perception, ESG affects how firms structure their financing.

These effects are not uniform. Governance and social factors tend to support borrowing, while environmental policies may introduce short-term constraints. Sectoral characteristics and crisis periods further shape these dynamics.

Taken together, these findings highlight the importance of integrating ESG considerations into both public policy and financial decision-making. They also suggest that sustainability and finance are increasingly interconnected, particularly in emerging economies undergoing structural transformation. These results reinforce the growing importance of integrating sustainability considerations into corporate finance frameworks, particularly in emerging and resource-dependent economies.

These findings also suggest that the integration of ESG into financial decision-making is not a temporary trend but part of a broader structural transformation of financial systems. As sustainability considerations become increasingly embedded in regulatory frameworks and investment strategies, their influence on corporate behavior is expected to grow.

In this context, sovereign ESG may play an increasingly central role in shaping not only financing conditions but also broader economic outcomes.

Future research could further explore these mechanisms across different institutional contexts and over longer time horizons.

Funding This work was supported by the Deanship of Scientific Research, Vice Presidency for Graduate Studies and Scientific Research, King Faisal University, Saudi Arabia, under Project Grant KFU261591.

References

Capelle-Blancard, G., Crifo, P., Diaye, M. A., Scholtens, B., & Oueghlissi, R. (2016). Environmental, social and governance (ESG) performance and sovereign bond spreads: an empirical analysis of OECD countries. Available at SSRN 2874262.

Crifo, P., Diaye, M. A., & Oueghlissi, R. (2017). The effect of countries' ESG ratings on their sovereign borrowing costs. *The Quarterly Review of Economics and Finance, 66*, 13–20.

Djankov, S., La Porta, R., Lopez-de-Silanes, F., & Shleifer, A. (2008). The law and economics of self-dealing. *Journal of Financial Economics, 88*(3), 430–465.

Friede, G., Busch, T., & Bassen, A. (2015). ESG and financial performance: Aggregated evidence from more than 2000 empirical studies. *Journal of Sustainable Finance & Investment, 5*(4), 210–233.

Goss, A., & Roberts, G. S. (2011). The impact of corporate social responsibility on the cost of bank loans. *Journal of Banking & Finance, 35*(7), 1794–1810.

Gratcheva, E., Emery, T., & Wang, D. (2021). Demystifying sovereign ESG. Available at SSRN 3854177.

Hübel, B., & Scholz, H. (2020). Integrating sustainability risks in asset management: The role of ESG exposures and ESG ratings. *Journal of Asset Management, 21*(1), 52–69.

Ioannou, I., & Serafeim, G. (2012). What drives corporate social performance? The role of nation-level institutions. *Journal of International Business Studies, 43*(9), 834–864.

Khan, M., Serafeim, G., & Yoon, A. (2016). Corporate sustainability: First evidence on materiality. *The Accounting Review, 91*(6), 1697–1724.

La Porta, R., Lopez-de-Silanes, F., Shleifer, A., & Vishny, R. (1998). Law and finance. *Journal of Political Economy, 106*(6), 1113–1155.

Levine, R. (2005). Finance and growth: Theory and evidence. *Handbook of Economic Growth, 1*, 865–934.

Myers, S. C. (1984). The capital structure puzzle. *The Journal of Finance, 39*(3), 575–592.

Ng, T. H., Lye, C. T., Chan, K. H., Lim, Y. Z., & Lim, Y. S. (2020). Sustainability in Asia: The roles of financial development in environmental, social and governance (ESG) performance. *Social Indicators Research, 150*, 17–44.

Rajan, R. G., & Zingales, L. (1995). What do we know about capital structure? Some evidence from international data. *The Journal of Finance, 50*(5), 1421–1460.

Wang, C., Ghadimi, P., Lim, M. K., & Tseng, M. L. (2019). A literature review of sustainable consumption and production: A comparative analysis in developed and developing economies. *Journal of Cleaner Production, 206*, 741–754.

Zhang, D., Zhao, Z., & Lau, C. K. M. (2022). Sovereign ESG and corporate investment: New insights from the United Kingdom. *Technological Forecasting and Social Change, 183*, 121899.

Green and Sustainability-Linked Bonds for Sovereigns and Cities

Ekaterina Pirozhkova

1 Introduction

The global market for labelled sovereign debt has undergone a structural transformation in less than a decade. When Poland issued the first sovereign green bond in December 2016, raising EUR 750 million for renewable energy and energy efficiency projects, such instruments were a novelty. By the end of 2025, the total issuance of sovereign sustainable bonds amounted to $92.2 billion (EF, 2026). Sovereign issuers collectively accounted for 12% of the green bond issues and 9% of the sustainability-linked bond issues (EF, 2026). Cities and subnational governments, though a smaller segment, have followed a parallel trajectory, with the Commonwealth of Massachusetts pioneering subnational green issuance as early as 2013 and municipalities across Latin America, Asia, and Africa increasingly exploring labelled instruments to finance urban resilience and low-carbon infrastructure (CAEO, 2021; MST, 2016).

This chapter examines how sovereigns and cities design and execute two distinct families of labelled instruments: green bonds (GBs) and sustainability-linked bonds (SLBs). Although both are commonly discussed under the umbrella of "sustainable finance," they rest on fundamentally different contractual logics and impose different institutional demands on issuers. Green bonds are use-of-proceeds instruments whose proceeds are ring-fenced for eligible environmental or climate projects; their credibility rests on the rigour of project selection, proceeds tracking, and impact reporting. Sustainability-linked bonds, by contrast, are performance-based instruments whose

The views expressed herein are solely those of the author and do not necessarily reflect the views of the South African Reserve Bank.

E. Pirozhkova (✉)
EM Normandie Business School, Métis Lab, Caen, France
e-mail: epirozhkova@em-normandie.fr

South African Reserve Bank, Pretoria, South Africa

H. Jarrar et al. (eds.), *Sovereign Sustainable Finance*, Sustainable Finance,
https://doi.org/10.1007/978-3-032-33239-4_3

financial characteristics—most commonly, the coupon—vary depending on whether the issuer achieves pre-defined sustainability targets. The credibility of SLBs rests on the materiality of the chosen indicators and the ambition of the targets against which they are measured.

The distinction between GBs and SLBs matters for practitioners. A finance ministry or city treasurer contemplating entry into the sustainability-labelled market must choose, first, between an instrument that disciplines how money is spent and one that disciplines what outcomes are achieved, and second, between a framework that demands robust public expenditure management and one that demands credible national statistics and policy commitment. Neither choice is inherently superior, as each carries distinct implications for governance architecture, investor communication, and post-issuance accountability.

This chapter synthesises current knowledge to provide actionable guidance for issuers of both types of instruments. Section 2 maps the normative and regulatory architecture governing these instruments. Section 3 addresses the design and execution of issuance frameworks, from key performance indicators (KPIs) selection and sustainability performance targets (SPTs) calibration to transaction mechanics and pricing. Section 4 examines post-issuance controls, including proceeds management, impact reporting, and the independent verification processes that make audit-ready governance possible. Section 5 illustrates these themes through four documented cases—France, Chile, Indonesia, and two contrasting subnational issuers that span advanced and emerging economies.

2 Market Architecture and Instrument Design

2.1 The ICMA Principles and the EU Green Bond Standard

The normative architecture of the sovereign labelled bond market is built on a set of voluntary principles published by the International Capital Market Association (ICMA) and, increasingly, on binding regulation in the European Union. The Green Bond Principles (GBP), first published in 2014 and most recently revised in 2025, define four core components for any instrument claiming the green label (ICMA, 2025):

1. use of proceeds, which must be allocated to eligible green project categories;
2. process for project evaluation and selection, requiring the issuer to communicate the environmental objectives and eligibility criteria;
3. management of proceeds, ensuring that allocated amounts are tracked and that unallocated funds are held in an appropriate manner;
4. reporting, mandating at minimum an annual allocation report and, where feasible, an impact report detailing the environmental outcomes achieved.

These components apply equally to sovereign and subnational issuers, though the specific mechanics of each component differ between a national treasury managing a consolidated fund and a city managing a project-level capital budget.

The Sustainability-Linked Bond Principles (SLBP), published in 2020 and revised in 2024, follow a parallel but distinct logic. An SLB is defined as any bond, for which financial and/or structural characteristics vary depending on whether the issuer achieves pre-defined sustainability performance objectives, measured through Key Performance Indicators (KPIs) assessed against Sustainability Performance Targets (SPTs) at specified observation dates. The five core components of the SLBP are (ICMA, 2024b):

1. selection of KPIs, which must be relevant, measurable, and externally verifiable;
2. calibration of SPTs, which must be ambitious relative to a credible baseline;
3. bond characteristics, specifying the financial mechanism triggered by SPT achievement or failure (typically, a coupon step-up);
4. reporting, requiring periodic disclosure of KPI performance;
5. verification, mandating independent assessment of KPI performance against SPTs at each observation date.

The step-up, or coupon ratchet, is the most common performance-based mechanism; a penalty of 25 basis points (bps) for non-compliance has become the market convention across both corporate and sovereign issuers (Vulturius et al., 2024).

A significant regulatory development took effect with the entry into force of the EU Green Bond Standard (EUGBS), established under Regulation (EU) 2023/2631 (EC, 2023). Unlike the ICMA GBP, the EUGBS is a binding regulation for bonds marketed as European Green Bonds within the EU. It requires full alignment of the use of proceeds with the EU Taxonomy for Sustainable Activities, engagement of a registered external reviewer,[1] and a prescribed disclosure and reporting template. The EUGBS simultaneously raises the level of transparency and creates a two-tier market: EU-regulated "European Green Bonds" and a continuing stream of GBP-aligned bonds, for which EUGBS compliance is not sought for (European Commission, 2023). For non-EU sovereign issuers, the EUGBS sets a benchmark that sophisticated European institutional investors may apply informally when assessing any sovereign green bond.

2.2 *Taxonomies and Eligible Categories*

The determination of which expenditures qualify as "green" depends on the taxonomy or classification system applied. For EU member states, the EU Taxonomy provides a technology-neutral, performance-based classification system covering

[1] A registered external reviewer is an entity supervised by the European Securities and Markets Authority (ESMA).

six environmental objectives: climate change mitigation, climate change adaptation, sustainable use and protection of water, circular economy, pollution prevention, and protection of biodiversity. For non-EU sovereigns, eligibility is typically determined by reference to ICMA's non-exhaustive list of eligible project categories, which includes renewable energy, energy efficiency, clean transportation, sustainable water management, climate change adaptation, sustainable management of natural resources and land use, and pollution prevention (ICMA, 2025).

The scope of eligible categories has gradually broadened. Canada's 2024 sovereign green bond framework became the first to include nuclear energy expenditure as an eligible category, following the EU Taxonomy's inclusion of nuclear under certain conditions (BNP Paribas Asset Management, 2024). Turkey's inaugural sovereign green bond in 2023 allocated nearly half its proceeds to sustainable management of natural resources and land use—one of the first emerging-market sovereigns to prioritise this category at scale (BNP Paribas Asset Management, 2024). Brazil's 2023 debut sustainable bond directed proceeds to deforestation control and biodiversity conservation, reflecting the importance of nature-based solutions in tropical biodiversity-rich economies (BNP Paribas Asset Management, 2024).

For subnational issuers, eligible categories tend to be more concentrated around urban infrastructure: energy-efficient buildings, clean public transport, sustainable water and wastewater systems, and climate-resilient urban infrastructure (ADB, 2024; Herrera, 2024). The alignment between a city's capital investment programme and available eligible categories is a practical first step in determining the feasibility of green issuance: a municipality whose capital budget is heavily weighted toward conventional road construction or social housing may find the eligible share of expenditure insufficient to justify the additional costs of labelling.

2.3 Structural Choices: Green Bonds versus Sustainability-Linked Bonds

From the perspective of a sovereign or city treasurer, the choice between a GB and an SLB reflects a fundamental trade-off between proceeds flexibility and outcome accountability. Green bonds ring-fence spending: the issuer commits to spending proceeds only on eligible expenditures, and credibility depends on demonstrating that this commitment has been honoured through post-issuance allocation reporting. SLBs ring-fence outcomes: the issuer commits to achieving a defined trajectory on one or more sustainability indicators, and credibility depends on demonstrating that performance has met or exceeded the target.

For advanced economy sovereigns with mature public expenditure management systems and large, diversified capital budgets, the green bond structure is typically simpler to implement: eligible expenditures are easily identified, proceeds tracking is integrated with existing treasury systems, and reporting draws on established government statistics. France, Germany, the United Kingdom, and Italy—the four largest sovereign green bond issuers globally—have all adopted the GB structure (CBI, 2025; Doronzo et al., 2021).

For emerging-market sovereigns, and particularly for those whose capital budgets are heavily constrained by fiscal adjustment programmes, SLBs offer an important advantage: proceeds are not ring-fenced, which means the instrument does not create a dual-track budget system or restrict the general use of borrowed funds. Chile's decision to issue the world's first sovereign SLB in March 2022 was motivated in part by this flexibility, as well as by the opportunity to link sovereign borrowing explicitly to the country's Nationally Determined Contribution (NDC) under the Paris Agreement (OECD et al., 2024; OIES, 2022). Uruguay subsequently issued SLBs tied to greenhouse gas intensity and afforestation targets, while the largest 2025 SLBs issues were carried by Republic of Slovenia and the Kingdom of Thailand (EF, 2026; OECD, 2024).

The SLB structure also introduces a risk that does not exist for GBs: the possibility that the issuer will fail to meet its SPT and will be required to pay the coupon step-up. For a sovereign with a large outstanding debt stock, even a 25-bp step-up on a single bond tranche may be fiscally manageable, but the reputational signal of non-compliance is significant. Conversely, if SPTs are set below the business-as-usual trajectory, the step-up mechanism is never triggered, and the instrument functions as a conventional bond with a sustainability label—a form of greenwashing that has been identified as a structural vulnerability of the SLB market (Flugge et al., 2021; Vulturius et al., 2024).

3 Framework Design and Execution

3.1 Developing the Sovereign or Municipal Green Bond Framework

The Green Bond Framework (GBF) or SLB Framework (SLBF) is the foundational governance document for any labelled issuance. It serves a dual purpose: internally, it codifies the issuer's eligibility criteria, project selection process, proceeds management procedures, and reporting commitments; externally, it is the primary document reviewed by the second-party opinion (SPO) provider and scrutinised by investors during due diligence. A well-designed framework communicates clearly that the issuer has done the institutional groundwork to deliver on its green commitments.

For sovereign issuers, the GBF must address the interface between capital market obligations and public budget law. This is non-trivial: in many jurisdictions, ring-fencing proceeds within a consolidated fund requires either a dedicated sub-account, a virtual tracking register maintained by the treasury, or a formal appropriation mechanism that earmarks specific budget lines for eligible expenditures. The alignment between the GBF's eligible categories and the government's existing budget classification system—whether a functional classification aligned with the classification of the functions of government or a programme-based classification—determines how easily eligible expenditures can be identified and tracked without creating a parallel accounting system. It is generally recommended that countries use their existing public financial management (PFM) infrastructure rather than building green-specific tracking systems from scratch. The World Bank's Green Bond Handbook (2022b) provides detailed guidance on designing the GBF.

For sovereign issuers in emerging markets, integration of the GBF with the country's Integrated National Financing Framework (INFF) or NDC investment plan can strengthen both instruments simultaneously. An INFF that has already mapped domestic climate financing needs, identified eligible public expenditures, and established monitoring and evaluation systems, provides the analytical backbone for a sovereign GBF at a relatively low additional cost (GISD Alliance, 2024). This integration also supports the narrative consistency, which is expected by sophisticated investors: a sovereign that issues green bonds should be able to demonstrate that those bonds are part of a coherent national financing strategy for climate action, not merely an opportunistic capital market transaction.

For municipal issuers, the design challenge is analogous but scaled differently. A city's GBF must map eligible expenditures within the municipal capital budget, align with the city's climate action plan or urban resilience strategy, and establish a project register that can be monitored at the individual project level. Best-practice municipal frameworks typically establish an inter-departmental green bond committee with representatives from the finance, environment, transport, and infrastructure departments, tasked with project selection, proceeds tracking, and reporting. Where institutional capacity is limited, this committee can be supported by technical assistance from multilateral development banks or development finance institutions (ADB, 2024).

3.2 KPI and SPT Design for Sustainability-Linked Instruments

For SLBs, KPI and SPT design is the most technically demanding and credibility-critical element. Flugge et al. (2021) identify necessary conditions regarding KPIs; they must be robust, properly interpreted, aligned with the country context, and credibly ambitious. ICMA's Illustrative KPI Registry, enhanced in 2023 with sovereign-specific indicators, provides a taxonomy of eligible metrics organised by sustainability theme—including absolute and intensity-based greenhouse gas metrics, renewable energy share, deforestation rate, and gender-equality indicators (ICMA, 2024b). SPTs must reflect genuine additionality beyond a business-as-usual trajectory, be consistent with science-based benchmarks, and produce observation-date milestones assessable before bond maturity. Chile's 2022 SLB provides the reference case: SPT 1 required absolute GHG emissions not to exceed 95 $MtCO_2e$ by 2030, directly aligned with its nationally determined contribution carbon budget;

SPT 2 required non-conventional renewable energy to reach 60 percent of installed capacity by 2032 (OIES, 2022). Both targets were independently verified by Sustainalytics and tied to a 25-bp step-up per KPI missed—a calibration that balances financial materiality against political manageability for sovereign issuers (Flugge et al., 2021).

3.3 External Reviews, Pricing, and Investor Engagement

External verification underpins market credibility. ICMA (2025) recommends appointing (an) external review provider(s) to assess via a pre-issuance review whether the GBs and/or GBF is aligned with the key components of the GBP. Post-issuance, it is recommended that an external auditor verifies the internal tracking and the allocation of funds from the GB proceeds. The second-party opinion, provided by an independent sustainability research firm assessing framework alignment with the relevant principles—is by far the most widely used for sovereign issuers. Evidence from Dorfleitner et al. (2022) and Baldi and Pandimiglio (2022) shows that SPOs from reputable providers are associated with a stronger green premium in primary markets, as investors price the credibility signal embedded in the external review. However, the lack of standardisation among SPO providers—each applying its own methodology with no regulatory oversight outside the EUGBS system—limits comparability across issuances (OECD, 2023). The EUGBS's registered reviewer system, with ESMA supervision, represents the most significant regulatory advance in addressing this gap and is expected to raise external review quality standards over time.[2]

Transaction execution adds an ESG-specific layer to conventional benchmark mechanics. Pre-deal investor engagement typically includes a dedicated sustainability roadshow targeting ESG-mandated institutional investors—pension funds, insurance companies, ESG-focused asset managers—whose participation broadens the investor base and signals credibility to conventional accounts. A central question for any issuer is whether the green label generates a pricing advantage. The empirical literature finds consistent but modest evidence of a positive greenium. Zerbib (2019) estimated an average of approximately 2 bps across government and corporate green bonds. For sovereign issuers specifically, Ando et al. (2023) found an average greenium of approximately 4 bps in advanced economies and 11 bps in emerging market economies, with the higher emerging-market premium attributed to stronger ESG investor demand relative to local supply and to the signalling value of sovereign commitment. More recent analysis of 2709 corporate green bonds issuance by 63 jurisdictions across 2012Q1–2022Q3 confirms similar magnitudes and finds that the greenium increases when climate policies are stronger (Cheng et al., 2024). The practical implication is that the direct fiscal benefit of sovereign green issuance is modest—often near zero once framework, external review, and reporting costs are

[2] ICMA's guidelines for external reviews and external review service mapping can be found at https://www.icmagroup.org/sustainable-finance/external-reviews/.

netted out—and the strategic rationale rests primarily on signalling credibility, developing domestic sustainable finance ecosystems, diversifying the investor base, and, for SLBs, creating a contractual accountability mechanism tied to climate targets (Caramichael & Rapp, 2022; Cheng et al., 2024).

4 Post-Issuance Controls and Integrity

4.1 Proceeds Management and Budget Integration

The credibility of a use-of-proceeds green bond depends ultimately on whether the issuer can demonstrate, post-issuance, that proceeds have been allocated to eligible expenditures as committed. This requires a proceeds management system that is both operationally functional and audit-ready, i.e. capable of producing a verifiable paper trail from the moment proceeds are received through disbursement to individual projects or budget lines.

For sovereign issuers, proceeds management must be integrated with the public financial management system. The World Bank (2022b) recommends one of three approaches depending on the issuer's public financial management capacity: (1) a dedicated sub-account within the treasury single account, into which green bond proceeds are deposited and from which disbursements to eligible expenditures are made; (2) a tracking register that maps existing budget appropriations to the green bond without creating a separate fund; or (3) a sub-portfolio approach in which a portfolio of eligible expenditures, already appropriated in the national budget, is designated ex-post as the green bond's use-of-proceeds pool. The sub-portfolio approach is the most common among advanced-economy sovereigns because it does not require changes to budget law and aligns naturally with the practice of issuing green bonds as tap or fungible instruments on existing benchmark lines (Agence France Trésor, 2023; Doronzo et al., 2021).

A critical risk in all three approaches stems from fungibility and the possibility that green bond proceeds displace rather than supplement eligible expenditures that would have occurred anyway. If a government issues USD 1 billion in green bonds and allocates the proceeds to renewable energy projects already appropriated in the budget, the additional climate impact (additionality) may be zero. The literature does not fully resolve this debate. Some note that the additionality question is ultimately unanswerable at the level of individual bond transactions, since government budgets are consolidated (Hardy, 2022; Lehmann & Martins, 2023). However, several mitigants are available: committing to expand the total budget allocation to eligible categories relative to a pre-issuance baseline; ringfencing green bond proceeds for new projects not in the original capital budget; and disclosing explicitly whether the green bond is financing new or previously appropriated expenditures (IMF, 2021; WB, 2022b).

For municipal issuers, project-level tracking is more tractable than for sovereigns because the capital budget is smaller and individual projects can be monitored at the asset level. Massachusetts' green bond programme illustrates the point: the Commonwealth was already tracking capital expenditures at the project level for internal due diligence purposes, and the additional burden of green bond reporting was minimal—consisting primarily of a reformatting of existing project data into investor-facing disclosure (CAEO, 2021). By contrast, cities in developing countries that lack robust project-level accounting systems may find that establishing a credible proceeds tracking mechanism requires significant upfront investment in PFM capacity (ADB, 2024; Herrera, 2024).

4.2 Impact Reporting and Harmonised Metrics

Annual impact reporting transforms the green bond from a financial instrument into a communication tool: it demonstrates to investors, policymakers, and citizens that the proceeds have generated measurable environmental benefits. ICMA's Harmonised Framework for Impact Reporting (2024a) provides sector-specific templates that define the recommended metrics for each eligible category, making it easier for issuers to produce reports that are comparable across the market. For example, the renewable energy template recommends reporting additional capacity of renewable energy constructed and rehabiliated (MW), annual generation (MWh), and annual greenhouse gas emissions reduced or avoided (tCO_2e); the energy efficiency template recommends reporting primary annual energy savings (MWh/year) and emissions avoided.

The distinction between allocation reporting and impact reporting is important and frequently conflated. Allocation reporting answers the question: which projects received green bond proceeds, and have all proceeds been fully allocated? Impact reporting answers the question: what environmental outcomes have been achieved as a result of those expenditures? The former is a compliance requirement; the latter is an accountability statement. Impact reporting is more demanding because it requires the issuer to construct counterfactual baselines, attribute outcomes to specific expenditures, and translate physical outcomes into standardised metrics. For sovereign issuers whose "projects" are in reality budget line items covering entire sectors, attribution is inherently difficult, and the most honest approach is to report at the programme level using sector-wide outcome indicators rather than claiming project-level attribution (OECD, 2023; WB, 2022b).

The quality of impact reporting has improved substantially since the early years of the green bond market, driven by ICMA harmonisation guidance and competitive pressure among issuers to produce credible, investor-grade reports. However, significant variation persists, particularly among emerging market issuers where national statistical systems may not produce the sector-specific data required for impact reporting on schedule. Multilateral support for capacity building in this area—including technical assistance for setting up monitoring and evaluation systems aligned with ICMA impact reporting templates—has become an important component of green bond readiness programmes (ADB, 2024; WB, 2022a).

4.3 Verification, Audit, and Greenwashing Risk

Post-issuance verification closes the accountability loop by providing independent confirmation that the issuer has honoured its green or sustainability-linked commitments. For green bonds, post-issuance verification typically covers two dimensions: allocation and impact. For SLBs, verification is more precisely defined: at each observation date, an independent verifier reviews the issuer's KPI data and confirms whether the SPT has been achieved, with the result determining whether the coupon step-up is triggered (ICMA, 2024b).

Greenwashing—the presentation of financial instruments as environmentally beneficial when the actual environmental impact is minimal or unverifiable, –poses the most significant reputational and regulatory risk for sovereign and subnational labelled bond programmes. In the green bond market, greenwashing risk arises primarily from three sources: first, overly broad eligibility criteria that include activities with marginal environmental benefit; second, inadequate project monitoring and impact measurement; and third, the absence of independent post-issuance verification. In the SLB market, structural greenwashing risks are different and in some respects more subtle: SPTs calibrated below the business-as-usual trajectory (so that compliance is guaranteed regardless of policy action); KPIs that measure activity rather than outcome (such as the number of green buildings certified rather than total building sector emissions); and observation dates structured so that the financial penalty for non-compliance applies only to the final coupon period before maturity, minimising the issuer's financial exposure (Fatica and Panzica, 2021; Flugge et al., 2021; Vulturius et al., 2024).

The regulatory response to greenwashing risk has been most developed in Europe. The EUGBS's requirement for registered external reviewers—entities supervised by European Securities and Markets Authority and subject to conduct-of-business requirements—represents a significant advance over the self-regulated second-party opinion (SPO) market. For non-EU issuers, the equivalent safeguard is the credibility and independence of the chosen SPO provider, the robustness of the verification methodology, and the comprehensiveness of the post-issuance audit trail. Issuers that invest in building an audit-ready documentation system—a project register with time-stamped expenditure records, independent measurement of physical outcomes, and a clear mapping between KPI calculations and their underlying data sources—are best positioned both to satisfy investor scrutiny and to withstand future regulatory requirements (OECD, 2023; WB, 2022b).

5 Country and City Case Studies

5.1 France: The Twin-Bond Model and Reporting Excellence

France's sovereign green bond programme, launched in January 2017 with a EUR 7 billion issuance, has established the reference model for advanced-economy sovereign governance. France adopted a "twin bond" structure: its Green OAT is issued as a fungible line mirroring an existing conventional OAT benchmark in all respects other than the use of proceeds, solving the secondary-market liquidity problem that deters institutional investors from holding non-fungible green bonds and enabling precise greenium measurement through daily secondary-market price comparison. Germany, the United Kingdom, and several other European sovereigns have since adopted the same design (Ando et al., 2023; Doronzo et al., 2021). France's annual Green OAT Impact Report—produced by the Agence France Trésor with independent third-party assurance—maps allocated expenditures across renewable energy, energy efficiency, sustainable transport, biodiversity, and climate adaptation, reporting physical outcomes using ICMA harmonised metrics (Agence France Trésor, 2023). The programme also demonstrates the market-development externalities: Cheng et al. (2024) find that a sovereign's debut green bond issuance spurs corporate green issuance in the same market, improves secondary-market liquidity, and strengthens demand for external reviews.

5.2 Chile: The First Sovereign SLB and NDC Alignment

Chile's March 2022 issuance of a USD 2 billion SLB—the first sovereign sustainability-linked bond—illustrates the strategic potential of performance-based instruments for middle-income emerging economies. The instrument was motivated by both the proceeds flexibility advantage and by the desire to link sovereign borrowing directly to the country's nationally determined contribution commitments (OIES, 2022). KPI 1 required absolute GHG emissions not to exceed 95 $MtCO_2e$ by 2030, aligned with Chile's NDC carbon budget; KPI 2 required non-conventional renewable energy capacity to reach 60 percent of total installed capacity by 2032. Both were independently verified by Sustainalytics, aligned with ICMA SLBP, and tied to a 25-bp coupon step-up per KPI missed. The transaction was more than three times oversubscribed, demonstrating strong international ESG investor demand. Chile subsequently issued further SLBs and innovated by adding a gender-equality KPI—the percentage of women in the national labour force—making it the first country to link sovereign bonds to both climate and social outcomes in the same instrument (OECD, 2024). The programme illustrates how SLBs can function as strategic communication tools for emerging-market sovereigns seeking to signal policy credibility and attract diversified international capital.

5.3 Indonesia: Emerging Market Green Sukuk

Indonesia's 2018 launch of the world's first sovereign green sukuk—an Islamic bond compliant with Shariah principles—exemplifies the adaptation of international green bond frameworks to a distinct domestic legal and cultural context. Developed in collaboration with the World Bank, the structure ensured both Shariah compliance and ICMA GBP alignment, demonstrating the framework's flexibility across legal systems (WB, 2022a). Between 2018 and end-2022, Indonesia raised approximately USD 6.9 billion through multiple green sukuk issuances, with oversubscription rates of 2.4–7.4 times, reflecting strong demand from an investor base combining ESG mandates and Islamic finance eligibility requirements (Owusu-Mante et al., 2025). Proceeds were allocated across renewable energy, sustainable transport, climate-resilient infrastructure, marine conservation, and waste management. Owusu-Mante et al. (2025) find, however, that the alignment between allocated expenditures and measurable environmental outcomes has been uneven, partly due to gaps in national monitoring and evaluation systems, what illustrates the broader challenge for emerging-market programmes: the financial instrument can be designed and executed relatively quickly, but the institutional infrastructure for credible impact reporting takes longer to build.

5.4 Subnational Issuers: Massachusetts and Developing-City Contrasts

The contrast between Massachusetts and the cities of Cape Town and Mexico City illustrates institutional readiness as the binding constraint on green municipal bond quality. Massachusetts completed its first USD 100 million green bond in 2013 and scaled the programme in 2014, producing in-house allocation and impact reports by reformatting pre-existing capital project monitoring data. Incremental costs were negligible, and the green label enabled a more compelling investor narrative—communicating specifically which infrastructure projects would be funded—that attracted both dedicated ESG accounts and a broader range of transparency-seeking institutional investors (CAEO, 2021).

Cape Town and Mexico City, both recognised sustainability leaders in their respective regions, faced substantially higher incremental costs: certification and SPO fees, external consultants for eligibility assessment and framework documentation, and additional legal costs that represented a meaningful proportion of total proceeds for inaugural or smaller issuances. In Mexico City, the measurement and attribution of environmental outcomes for water infrastructure projects also proved difficult to document rigorously (Herrera, 2024). Taken together, these cases establish that when public financial management and project-monitoring infrastructure is already robust, the marginal cost of green labelling is low and the strategic benefits are real; where that infrastructure is lacking, multilateral technical assistance, partial credit guarantees, and co-issuance structures can play a decisive role in enabling first-time issuers (ADB, 2024; World Bank, 2022a).

A cross-cutting lesson from all four cases is the importance of investor communication strategy alongside technical compliance. France's regular investor briefings and high-quality impact reports have created a loyal green investor base that reliably participates in subsequent tap issuances at favourable prices. Chile's proactive engagement with ESG rating agencies and international investors before each SLB contributed to strong oversubscription. Indonesia's green sukuk benefited from the Shariah-compliance certification as a dual-access signal. And Massachusetts' straightforward, in-house reporting style proved that investor communication need not be expensive to be effective. In each case, the credibility of the communication rested ultimately on the quality of the underlying governance, demonstrating that investor relations in the labelled bond market is a consequence of institutional integrity.

6 Conclusion

Green bonds and sustainability-linked bonds have matured from innovative instruments to standard components of the sovereign debt management toolkit. The normative infrastructure—ICMA's Green Bond Principles and Sustainability-Linked Bond Principles, the EU Green Bond Standard, the Climate Bonds Initiative's certification scheme—provides stable frameworks within which even first-time issuers can structure credible transactions. The growth of the SPO industry and the gradual harmonisation of impact reporting metrics have reduced information asymmetry between issuers and investors, while the EU's regulatory intervention through the EUGBS is raising the level of transparency and verification in Europe and, indirectly, setting benchmarks for global practice.

Four conclusions emerge from the analysis in this chapter. First, instrument choice should be driven by institutional fit, not financial opportunism. The green bond structure is better suited to issuers with mature public financial management systems that can support ring-fenced proceeds tracking and project-level impact reporting. The SLB structure is better suited to issuers—particularly emerging-market sovereigns, who seek proceeds flexibility and wish to create a contractual accountability mechanism tied to national climate or sustainable development targets. Neither instrument is appropriate for an issuer that lacks the institutional infrastructure to deliver on its commitments.

Second, pre-issuance design quality is the primary determinant of programme credibility. The robustness of the GBF or SLBF—specifically, the integration with budget classification systems, the rigour of KPI and SPT design, and the credibility of the chosen external reviewer—establishes the baseline of investor confidence that subsequent reporting and verification must maintain. Shortcuts at the design stage are difficult and costly to remedy after issuance.

Third, the greenium is real but modest. The weight of evidence suggests that sovereign green bonds generate a yield advantage of 2–13 basis points depending on the issuer's market context, with larger premia in emerging markets, where ESG supply-demand imbalances are more pronounced. This advantage is insufficient on its own to justify the additional costs of labelled issuance for most advanced-economy sovereigns. The strategic rationale rests on complementary benefits: signalling credibility, developing domestic sustainable finance ecosystems, diversifying the investor base, and for SLBs, creating a public accountability mechanism tied to climate commitments.

Fourth, institutional capacity is the binding constraint for subnational and emerging-market issuers. The contrast between Massachusetts and Cape Town or Mexico City is not primarily a contrast in financial market access or sustainability ambition; it is a contrast in the public financial management and monitoring infrastructure that determines whether green labelling can be delivered at low incremental cost or requires substantial capacity investment. Multilateral development banks and development finance institutions have a significant role to play in bridging this gap, through technical assistance programmes, partial credit guarantees that reduce borrowing costs for first-time issuers, and co-issuance structures that allow smaller municipalities to pool eligible expenditures.

Looking ahead, the convergence of rising regulatory standards, growing ESG investor mandates, and deepening national taxonomy frameworks is creating conditions, in which credible green and SLB programmes will become progressively easier to execute and more valuable to maintain. Sovereigns and cities that invest in the institutional foundations, such as integrated budget classification, systematic project monitoring, transparent impact reporting, will be well positioned to meet these rising expectations while advancing their climate and sustainable development objectives.

References

Agence France Trésor. (2023). *Green OAT: 2023 impact report. Ministère de l'Économie, des Finances et de la Souveraineté industrielle et numérique.* https://www.aft.gouv.fr

Ando, S., Fu, C., Roch, F., & Wiriadinata, U. (2023). *How large is the sovereign greenium?* (IMF Working Paper WP/23/80). International Monetary Fund. https://doi.org/10.5089/9798400235160.001

Asian Development Bank (ADB). (2024). *Accelerating green bonds for municipalities in Southeast Asia.* ADB. https://doi.org/10.22617/TCS240472

Baldi, F., & Pandimiglio, A. (2022). The role of ESG scoring and greenwashing risk in explaining the yields of green bonds: A conceptual framework and an econometric analysis. *Global Finance Journal, 52*, 100711. https://doi.org/10.1016/j.gfj.2022.100711

BNP Paribas Asset Management. (2024). *Sovereign green bonds: Bigger, stronger, and more diverse in 2024.* https://www.bnpparibas-am.com/en/portfolio-perspectives/sovereign-green-bonds-bigger-stronger-and-more-diverse-in-2024/

Caramichael, J., & Rapp, A. (2022). *The green corporate bond issuance premium* (International Finance Discussion Paper 1346). Board of Governors of the Federal Reserve System. https://doi.org/10.17016/IFDP.2022.1346

Cheng, G., Ehlers, T., Packer, F., & Xiao, Y. (2024). *Sovereign green bonds: A catalyst for sustainable debt market development?* (IMF Working Paper WP/24/120). International Monetary Fund. https://www.imf.org/-/media/files/publications/wp/2024/english/wpiea2024120-print-pdf.pdf

Climate Action Economic Opportunities (CAEO). (2021). *Exploring economic and enterprise opportunities from climate action.* https://lgiu.org/wp-content/uploads/2021/11/Economic-opportunities-from-climate-action-LGIU-CARO-report-2021.pdf

Climate Bonds Initiative (CBI). (2025). *Sustainable debt: Global state of the market 2024.* CBI. https://www.climatebonds.net/files/documents/publications/Climate-Bonds_Sustainable_Debt_2024_Global-State-of-the-Market_24-Sep-2025.pdf

Dorfleitner, G., Utz, S., & Zhang, R. (2022). The pricing of green bonds: External reviews and the shades of green. *Review of Managerial Science, 16*, 797–834. https://doi.org/10.1007/s11846-021-00458-9

Doronzo, R., Siracusa, V., & Antonelli, S. (2021). *Green bonds: The sovereign issuers' perspective* (Bank of Italy Markets, Infrastructures, Payment Systems Working Paper No. 3). https://www.bancaditalia.it/pubblicazioni/mercati-infrastrutture-e-sistemi-di-pagamento/approfondimenti/2021-003/index.html?com.dotmarketing.htmlpage.language=1

Environmental Finance (EF). (2026). *Sustainable bonds insight 2026.* https://www.environmental-finance.com/content/downloads/

European Commission (EC). (2023). *Regulation (EU) 2023/2631 on European green bonds and optional disclosures for bonds marketed as environmentally sustainable and for sustainability-linked bonds.* Official Journal of the European Union. https://eur-lex.europa.eu

Fatica, S., & Panzica, R. (2021). Green bonds as a tool against climate change? *Business Strategy and the Environment, 30*(5), 2688–2701. https://doi.org/10.1002/bse.2771

Flugge, M. L., Mok, R. C. K., & Stewart, F. E. (2021). *Striking the right note: Key performance indicators for sovereign sustainability-linked bonds.* World Bank Group. https://documents1.worldbank.org/curated/en/935681641463424672/pdf/Striking-the-Right-Note-Key-Performance-Indicators-for-Sovereign-Sustainability-Linked-Bonds.pdf

Global Investors for Sustainable Development (GISD) Alliance. (2024). *Guidance on sovereign SDG bonds for countries and investors.* GISD. https://www.gisdalliance.org/sites/default/files/2024-04/GISD%20Alliance%20Guidance%20on%20Sovereign%20SDG%20Bonds-FINAL.pdf

Hardy, D. (2022). *Alternatives in the design of sovereign green bonds. wiiw Policy Note and Report 62.* The Vienna Institute for International Economic Studies. https://wiiw.ac.at/alternatives-in-the-design-of-sovereign-green-bonds-dlp-6412.pdf

Herrera, H. (2024). The proliferation of municipal green bonds in Africa and Latin America: The need for a climate justice approach. *Environment & Urbanization, 36*(1), 147–172. https://doi.org/10.1177/09562478241230290

International Capital Market Association (ICMA). (2024a). *Handbook: Harmonised framework for impact reporting* (June 2024 edition). ICMA. https://www.icmagroup.org/assets/documents/Sustainable-finance/2024-updates/Handbook-Harmonised-Framework-for-Impact-Reporting-June-2024.pdf

International Capital Market Association (ICMA). (2024b). *Sustainability-linked bond principles: Voluntary process guidelines* (June 2024 edition). ICMA. https://www.icmagroup.org/sustainable-finance/the-principles-guidelines-and-handbooks/sustainability-linked-bond-principles-slbp/

International Capital Market Association (ICMA). (2025). *Green bond principles: Voluntary process guidelines for issuing green bonds* (June 2025 edition). ICMA. https://www.icmagroup.org/sustainable-finance/the-principles-guidelines-and-handbooks/green-bond-principles-gbp/

International Monetary Fund (IMF). (2021). *Fiscal policies to address climate change in Asia and the pacific.* IMF. https://www.imf.org/-/media/files/publications/dp/2021/english/fpaccapea.pdf

Lehmann, A., & Martins, C. (2023). *The potential of sovereign sustainability-linked bonds in the drive for net-zero.* Policy Brief 07/2023, Bruegel. https://www.bruegel.org/policy-brief/potential-sovereign-sustainability-linked-bonds-drive-net-zero

Massachusetts State Treasury (MST). (2016). *Mass green bonds* (Final Investor Impact Report, June 2016). https://www.massbondholder.com/sites/default/files/2020-08/2017-green-bond-report.pdf

Organisation for Economic Co-operation and Development (OECD). (2023). *Report on green, social and sustainability bonds issued by members and non-members of the OECD* (DAF/CMF/AS(2023)3/REV2). OECD. https://one.oecd.org/document/DAF/CMF/AS(2023)3/REV2/en/pdf

Organisation for Economic Co-operation and Development (OECD). (2024). *The surge of green, social, sustainability and sustainability-linked (GSSS) bonds in Latin America and the Caribbean* (OECD Development Policy Papers No. 56). https://doi.org/10.1787/f1c893a3-en.

Owusu-Mante, S., Hua-Ke Chi, K., & Gopalakrishnan, T. (2025). Sovereign green bonds in practice: Effectiveness as a climate finance instrument in emerging economies. *Climate Policy*. https://doi.org/10.1080/14693062.2025.2586572

Oxford Institute for Energy Studies (OIES). (2022). *Sustainability-linked bonds and their role in the energy transition* (OIES Paper ET18). https://www.oxfordenergy.org/publications/sustainability-linked-bonds-and-their-role-in-the-energy-transition/

Vulturius, G., Maltais, A., & Forsbacka, K. (2024). Sustainability-linked bonds: Their potential to promote issuers' transition to net-zero emissions and future research directions. *Journal of Sustainable Finance & Investment, 14*(1), 116–127. https://doi.org/10.1080/20430795.2022.2040943

World Bank (WB). (2022a). *Sovereign green, social and sustainability bonds: Unlocking the potential in emerging markets*. World Bank Group. https://thedocs.worldbank.org/en/doc/4de3839b85c57eb958dd207fad132f8e-0340012022/original/WB-GSS-Bonds-Survey-Report.pdf

World Bank (WB). (2022b). *Green bond handbook: A step-by-step guide to issuing a green bond*. World Bank Group. http://documents.worldbank.org/curated/en/099854406222232086

Zerbib, O. D. (2019). The effect of pro-environmental preferences on bond prices: Evidence from green bonds. *Journal of Banking & Finance, 98*, 39–60. https://doi.org/10.1016/j.jbankfin.2018.10.012

Budget Tagging and Climate-Informed Public Financial Management

Charbel Salloum

1 Introduction

Climate policy becomes fiscally meaningful only when it enters the machinery of public finance. Nationally determined contributions, adaptation plans, and long-term strategies may state direction, yet annual budgets still decide which ministries receive resources, which projects move forward, and which risks are disclosed. That is why climate-aware budgeting has shifted from a specialist topic to a core concern for ministries of finance, auditors, and debt managers (Battersby et al., 2022; OECD, 2024).

Budget tagging has become one of the most visible tools in this shift. It promises a practical bridge between climate objectives and public expenditure by classifying budget items according to their climate relevance. The attraction is obvious. Tagging is administratively more feasible than a full overhaul of budget systems, and it can produce an early map of where climate-related resources are flowing. The wider international uptake is equally clear. In the OECD survey underpinning the 2024 stocktake, 16 of 24 countries reported climate and environmental matters through the budget process in 2022, while tagging and related reporting tools continued to spread across advanced and emerging economies (OECD, 2024).

Even so, mature practice now points to a harder truth. Tagging is useful, but it is not a self-executing reform. A government can tag thousands of budget lines and still fail to improve project selection, medium-term prioritization, procurement, or fiscal risk management. World Bank and IDB reviews make this point repeatedly. Classification systems create visibility, but visibility alone does not guarantee better decisions, stronger accountability, or credible climate outcomes (Pizarro et al., 2021; World Bank, 2021, 2025).

C. Salloum (✉)
Métis Lab, EM Normandie Business School, Clichy, France
e-mail: csalloum@em-normandie.fr

H. Jarrar et al. (eds.), *Sovereign Sustainable Finance*, Sustainable Finance,
https://doi.org/10.1007/978-3-032-33239-4_4

This chapter, therefore, treats budget tagging not as an end state but as an entry point into climate-informed public financial management. The argument proceeds in four steps. First, it explains why tagging matters and where its limits begin. Second, it identifies the design choices that distinguish a credible tagging architecture from a symbolic one. Third, it shows how tagging should connect to the broader budget cycle, including public investment management and fiscal risk analysis. Fourth, it outlines a practical reform sequence for governments that want climate-responsive budgeting to influence real budget choices rather than remain a reporting annex.

2 Why Climate Budget Tagging Matters, and Why It Is Not Enough

Climate budget tagging became popular because it solves a real administrative problem. Many governments know they are spending on mitigation, adaptation, or related enabling measures, but cannot identify the amounts with confidence inside ordinary budget classification systems. Tagging helps create that first layer of visibility. It enables ministries of finance to track spending patterns, support climate finance reporting, inform development partners, and communicate more clearly with legislatures and the public (OECD, 2021; World Bank, 2021, 2025).

At its best, tagging also improves internal coordination. Once line ministries are required to identify climate-relevant budget lines, climate policy will no longer sit solely with environmental agencies. Transport, agriculture, water, housing, energy, and local government portfolios all become part of a shared fiscal conversation. This cross-government effect is one reason the IMF frames green PFM as a mainstreaming exercise rather than a parallel climate bureaucracy. The point is to adapt existing budget institutions rather than bypass them (Battersby et al., 2022).

Still, the evidence base now shows that tagging can disappoint when its objective is vague. Some systems are designed primarily for external reporting. Others aim to inform budget prioritization, public investment management, or expenditure review. Those aims are not identical. A methodology that is good enough for communication may be too coarse for decision-making. OECD guidance is explicit on this point. A tagging exercise should start by answering a precise management question, because the design of categories, coverage, and validation depends on the use case (OECD, 2021).

A second limitation concerns classification itself. Many early exercises focused only on expenditures expected to support climate goals. That creates a partial picture. It tells policymakers what is favorable, but not which measures may be environmentally harmful, contradictory, or fiscally inefficient. The more advanced green budgeting literature, therefore, argues for moving beyond positive-only identification toward a fuller treatment of favorable, neutral, and unfavorable measures, especially where subsidies, tax expenditures, or infrastructure choices may work against climate objectives (Pojar, 2023; Postic & Cardot, 2021).

Another difficulty is that tags often describe intention rather than effect. A project may be labeled as adaptation-relevant because it contains resilience language, while the underlying design remains weak. Likewise, a mitigation tag does not prove cost-effectiveness or emissions impact. This is why the World Bank and IDB work stresses compatibility with statistical systems, project appraisal methods, and expenditure control procedures. Classification must be connected to evidence if it is to guide policy rather than simply decorate the budget (Pizarro et al., 2021; World Bank, 2021, 2025).

For that reason, mature practice no longer treats tagging as the whole reform. It treats tagging as one tool within green budgeting and climate-responsive PFM. Governments that use it well generally pair it with clearer budget circulars, stronger public investment screening, medium-term planning, and more disciplined reporting. Governments that use it poorly often obtain a climate tally but little change in the quality of budget decisions (Battersby et al., 2022; OECD, 2024) (Table 1).

Table 1 Core design choices in climate budget tagging

Design choice	What strong practice looks like	What goes wrong when it is weak
Objective of tagging	Method answers a concrete management question such as reporting, prioritisation, or expenditure review	Tagging is launched as a general climate signal and produces numbers that no budget actor actually uses
Coverage and categories	Mitigation, adaptation, enabling and, where relevant, unfavorable measures are defined transparently	Positive spending is counted while contradictory measures remain invisible
Institutional ownership	Finance ministry leads, line ministries propose, and a challenge function validates classifications	Line ministries self-tag without scrutiny or central consistency checks
Budget integration	Tags appear in circulars, budget documentation, execution reports, and public investment processes	Tags are added late in the cycle and survive only as an annex
Data and review	FMIS fields, guidance notes, training, and revision rules support annual updating	Manual spreadsheets, inconsistent interpretation, and no audit trail undermine reliability

Source Author's own elaboration based on OECD (2021, 2024), World Bank (2021, 2025), Pizarro et al. (2021), and Battersby et al. (2022)

3 Designing a Credible Tagging Architecture

A credible tagging architecture begins with scope and definitions. Governments need to decide whether the exercise will cover only expenditures, or also revenues and tax expenditures; whether it will classify only climate items or broader environmental measures; and whether it will differentiate between mitigation, adaptation, and enabling expenditures. The World Bank review of international practice shows substantial variation across these design choices, which is why comparability across countries is often more limited than headline summaries imply (World Bank, 2021, 2025).

Classification rules matter just as much as coverage. Some countries rely on binary tags, but many have moved toward graded systems because climate relevance is rarely all-or-nothing. Weighted or scaled approaches can be more informative when programs have mixed objectives, although they also require stronger guidance and more administrative judgment. OECD and IDB guidance both emphasize that a methodology must be sufficiently understandable for line ministries to apply consistently, while remaining precise enough to avoid inflating climate numbers through over-tagging (OECD, 2021; Pizarro et al., 2021).

Ownership is another decisive issue. Ministries of finance typically need to lead the methodology because tagging affects budget preparation, budget classification, and reporting. Yet finance ministries should not work alone. Sector ministries possess the technical expertise needed to assess whether a line item genuinely contributes to mitigation or adaptation. Better systems, therefore, combine central guidance with line-ministry self-assessment and some form of challenge or validation. Without that challenge function, tagging can quickly become optimistic self-reporting.

The operational unit of tagging should also be chosen with care. Tagging at the level of whole programs may be administratively simple, but it can conceal mixed spending within broad envelopes. Tagging highly granular budget lines can improve accuracy, although at the cost of a heavier administrative burden. The right compromise depends on the budget structure and the quality of financial management information systems. This is why recent technical assistance has often recommended phased implementation rather than immediate full coverage (International Monetary Fund, Fiscal Affairs Department, 2024d).

Methodological credibility also depends on documented exclusions, revision rules, and treatment of brown measures. The literature has become more skeptical of methodologies that publish only favorable numbers and leave harmful measures off stage. European and OECD practice increasingly points toward fuller budget assessments that acknowledge trade-offs and policy inconsistencies rather than presenting climate budgeting as a one-sided inventory of positive action (OECD, 2024; Pojar, 2023; Postic & Cardot, 2021).

Finally, systems need quality assurance. Clear manuals, reviewer notes, training cycles, and auditable documentation help transform tagging from a one-off exercise into a repeatable budget process. In practical terms, a strong tagging architecture is not the one with the most categories. It is the one that produces classifications that ministries can defend, update, and use in subsequent budget decisions.

4 Embedding Tagging in the Wider PFM Cycle

Once tagging is designed, the central reform question becomes whether it changes the wider budget cycle. Green PFM guidance is clear that climate relevance should be incorporated before the annual budget is finalized and maintained after appropriations are approved. In other words, tagging works best when it is linked to strategic planning, medium-term expenditure frameworks, public investment management, budget execution, reporting, and oversight rather than treated as a downstream accounting exercise (Battersby et al., 2022; UNDP, 2021).

The budget circular is one of the most important points of connection. When the circular requires ministries to explain the climate relevance of proposed spending, the discussion starts early enough to influence allocations and project selection. When tagging happens only after ceilings are fixed, it mainly documents decisions already taken. This distinction helps explain why some governments obtain more value from tagging than others. Timing determines whether climate information informs budget choices or merely describes them.

Public investment management is especially important here because many of the largest climate consequences sit in capital spending. IMF climate public investment assessments in Kenya, Egypt, and Cabo Verde all underline a similar issue. Governments may have improved planning and coordination, yet still need to integrate climate considerations more systematically into appraisal, selection, budgeting, portfolio management, and fiscal risk analysis (International Monetary Fund, Fiscal Affairs Department, 2024a, 2024b, 2024c). Tagging can support this process, but only if project documents, appraisal templates, and selection criteria are aligned with the tagging methodology.

Execution and reporting are equally important. A budget tag assigned during preparation should be traceable during budget execution and visible in year-end reporting to support accountability. OECD survey evidence shows progress in this direction, with more countries using budget reports, performance indicators, and dashboards to disclose climate-related budget information. Yet the same evidence suggests that monitoring remains weaker than the front-end design of green budgeting tools in many jurisdictions (OECD, 2024).

Country experience illustrates why implementation discipline matters. The Uganda PFM climate assessment noted that climate budget tagging had been introduced, but was not yet producing fully reliable estimates of resources used for climate objectives, and therefore required stronger methodology and training. Seychelles, by contrast, received advice to phase its framework carefully, embed it in the budget cycle, and treat tagging as a process for identifying, measuring, and monitoring climate-relevant spending over time rather than as a one-year classification event (International Monetary Fund, Fiscal Affairs Department, 2024d, 2024e).

Climate-informed PFM also extends beyond expenditure coding. Procurement, asset management, and fiscal risk management all matter. A country may classify adaptation expenditure correctly but still fail to maintain climate-vulnerable assets, disclose contingent liabilities, or procure resilient infrastructure. Green PFM is therefore broader than tagging by design. Tagging provides the map, but the budget system still needs to steer the vehicle (Battersby et al., 2022; PEFA Secretariat, 2024).

5 Sequencing Reforms for Decision Usefulness and Accountability

Sequencing is where many otherwise sensible reforms struggle. Governments often feel pressure to produce climate numbers quickly, especially when reporting to development partners or signaling commitment to markets. Yet experience suggests that a narrow but credible first phase is usually better than an ambitious methodology that collapses under administrative strain. OECD, IMF, and PEFA materials all point to a phased reform approach that begins with institutional readiness, clear rules, and a manageable pilot scope (OECD, 2021; PEFA Secretariat, 2024).

A practical first phase usually includes four steps. The first is to define the policy objective and classification logic. The second is to issue a budget circular and guidance note that line ministries can apply consistently. The third is to test the methodology in a limited set of ministries or sectors where climate relevance is already visible, such as transport, water, energy, agriculture, or disaster risk management. The fourth is to validate results and refine the system before expanding coverage. This may sound unglamorous, yet it is precisely the administrative realism that separates working systems from symbolic ones.

The second phase should deepen the use of decisions. That means linking tags to spending reviews, project appraisal, medium-term budgeting, and fiscal risk reporting. It may also mean extending coverage to tax expenditures and environmentally harmful measures, where politically feasible. The Pojar analysis of European practice is helpful here because it shows that green budgeting matures when it is embedded in standard budgetary institutions rather than treated as a side exercise run solely by climate specialists (Pojar, 2023).

Independent scrutiny is part of the same maturation process. Legislatures, supreme audit institutions, and public financial accountability frameworks can help test whether climate-related budget information is reliable and policy-relevant. The first edition of PEFA Climate was created precisely to provide a more systematic lens on how well PFM systems support climate policy through the full budget cycle. Used well, such diagnostics can prevent governments from mistaking the existence of a tag for the existence of climate-responsive budgeting (PEFA Secretariat, 2024).

Training and institutional memory deserve more attention than they usually receive. Tagging methodologies often fail not because the concept is flawed, but because staff rotate, guidance is thin, or line ministries interpret categories differently from year to year. IMF country assessments in Uganda, Kenya, Egypt, and Cabo Verde repeatedly return to this administrative core. Better climate budgeting depends on manuals, capacity, and repeatable routines as much as on high-level strategy (International Monetary Fund, Fiscal Affairs Department, 2024a, 2024b, 2024c, 2024e).

The deeper lesson is that climate-informed PFM should make budgets more decision-ready, not simply more climate-literate. A budget system that can distinguish favorable from unfavorable measures, connect capital spending to climate objectives, and disclose climate-related fiscal risks is more useful than one that produces a large headline number without explaining how that number affected prioritization. Credibility grows when climate information changes decisions, not when it merely fills tables.

6 Conclusion

Climate budget tagging has moved from experimental practice to a mainstream entry point for climate-responsive budgeting. That growth is justified. Tagging can create a common fiscal language for climate policy, improve transparency, and support coordination across ministries. In many countries, it is the first practical step that allows finance ministries to see where climate-relevant resources are being allocated.

But the evidence reviewed here also shows where the frontier now lies. Mature practice is no longer about whether a government has a tag. It is about whether that tag is connected to ceilings, appraisals, procurement, execution, reporting, and oversight. Climate-informed PFM begins when classification starts to influence choice.

For governments seeking durable reform, the priorities are straightforward. Define the objective clearly. Keep the methodology defensible. Build validation and documentation into the process. Link tags to the wider budget and public investment cycle. Expand only when institutional capacity can sustain the next step. That is how climate budgeting becomes more than a technical annex and begins to function as part of core fiscal governance.

References

Battersby, B., Gonguet, F., Wendling, C. P., Charaoui, J., & Petrie, M. (2022). *How to make the management of public finances climate-sensitive—"green PFM"* (IMF How To Notes, 2022(006)). https://doi.org/10.5089/9798400221682.061

International Monetary Fund, Fiscal Affairs Department. (2024a). *Arab Republic of Egypt: Technical assistance report—climate public investment management assessment* (Technical Assistance Report No. 2024/029). https://doi.org/10.5089/9798400271960.019

International Monetary Fund, Fiscal Affairs Department. (2024b). *Cabo verde: Technical assistance report—climate public investment management assessment* (C-PIMA) (Technical Assistance Report No. 2024/034). https://doi.org/10.5089/9798400272479.019

International Monetary Fund, Fiscal Affairs Department. (2024c). *Kenya: Technical assistance report—climate module of the public investment management assessment* (Technical Assistance Report No. 2024/009). https://doi.org/10.5089/9798400268342.019

International Monetary Fund, Fiscal Affairs Department. (2024d). *Seychelles: Technical assistance report—developing a framework for climate budget tagging* (Technical Assistance Report No. 2024/082). https://doi.org/10.5089/9798400288562.019

International Monetary Fund, Fiscal Affairs Department. (2024e). *Uganda: Technical assistance report—PFM climate assessment: Public investment and fiscal risks management* (Technical Assistance Report No. 2024/012). https://doi.org/10.5089/9798400268168.019

OECD. (2021). *Green budget tagging: Introductory guidance and principles*. OECD Publishing. https://doi.org/10.1787/fe7bfcc4-en

OECD. (2024). *Green budgeting in OECD countries 2024*. OECD Publishing. https://doi.org/10.1787/9aea61f0-en

PEFA Secretariat. (2024). *Supplementary framework for assessing climate-responsive public financial management (PEFA climate)* (1st ed.). PEFA.

Pizarro, R., Delgado, R., Eguino, H., & Pereira, A. L. (2021). *Climate change public budget tagging: Connections across financial and environmental classification systems*. Inter-American Development Bank. https://doi.org/10.18235/0003021

Pojar, S. (2023). *How green budgeting is embedded in national budget processes* (European Economy Discussion Paper No. 196). Directorate-General for Economic and Financial Affairs, European Commission.

Postic, S., & Cardot, F. (2021). *Environmental budget tagging: The good, the bad and the unclear*. Institute for Climate Economics.

United Nations Development Programme. (2021). *Budgeting for climate change: A guidance note for governments to integrate climate change into budgeting, with a focus on medium-term budgets*. UNDP.

World Bank. (2021). *Climate change budget tagging: A review of international experience. Equitable growth, finance and institutions insight*. World Bank. https://doi.org/10.1596/35174

World Bank. (2025). *Climate-smart public financial management: MENA conference proceedings*. World Bank.

Disaster and State-Contingent Clauses in Sovereign Debt: Trigger Design, Pricing, and Documentation

Houanti L'Hocine

1 Introduction

Climate-related disasters punish sovereigns twice. The first blow is physical and economic: damaged infrastructure, lower output, disrupted revenues, and urgent humanitarian spending. The second blow is financial: the shock arrives just when debt service still has to be paid and new borrowing becomes more expensive or harder to arrange. State-contingent clauses try to correct that timing problem by making at least part of the debt contract respond automatically to a clearly defined shock (Ando et al., 2022; Cohen et al., 2020).

The attraction is obvious, but the design challenge is equally obvious. A clause that is too vague will be discounted by investors or abused by issuers. A clause that is too narrow may never be triggered when it is most needed. What matters here is not the slogan of resilience, but the contract architecture behind it. This chapter focuses on mature lessons that now emerge from the literature and from practice: the economic case for contingency, the mechanics of trigger design, the pricing and legal questions that shape credibility, and the conditions under which these clauses genuinely improve sovereign resilience.

The case for these clauses rests on timing. After a severe disaster, the state needs liquidity before it needs a full macroeconomic reassessment. Relief spending, restoration of essential services, and early reconstruction all have to start quickly, yet tax receipts often fall at the same moment. In highly exposed sovereigns, reserve buffers and insurance can help, but they are not always large enough or fast enough to absorb the entire gap. This is why state-contingent clauses deserve attention as part of the broader resilience stack. They are not grants and they are not debt forgiveness. They are contractual tools for shifting cash flow across time when the shock is clearly

H. L'Hocine (✉)
Métis Lab, EM Normandie Business School, Clichy, France
e-mail: lhouanti@em-normandie.fr

H. Jarrar et al. (eds.), *Sovereign Sustainable Finance*, Sustainable Finance,
https://doi.org/10.1007/978-3-032-33239-4_5

exogenous and the liquidity need is immediate (Ando et al., 2022; Cohen et al., 2020; World Bank, 2024).

There is also an institutional advantage that matters in practice. When payment relief is defined ex ante in the contract, governments do not have to reopen negotiations in the middle of an emergency simply to secure short-term breathing space. That reduces delay, political friction, and the stigma that can come from approaching markets or official creditors after a disaster has already damaged fiscal credibility. Even so, the benefit only appears when the clause is specific enough to be trusted. General language about resilience does not help much. Investors and policymakers need to know exactly what event triggers the clause, how payments are rescheduled, who verifies the trigger, and what happens next in the debt-service calendar (Guerson, 2021; Knight and Northfield, 2020; World Bank and International Monetary Fund, 2014).

2 Why State Contingency Matters for Sovereign Resilience

The economic case for contingency starts with liquidity, not generosity. After a disaster, governments must fund relief, restore critical services, and begin reconstruction before growth and revenue recover. If debt service continues on its original path, the sovereign may have to cut essential spending, draw down already thin reserves, or seek expensive bridge financing. A temporary and predefined deferral can soften that cliff edge and reduce the risk that a liquidity shock turns into a deeper fiscal and social crisis (Guerson, 2021; Igan et al., 2021).

Caribbean practice gives this logic concrete form. Grenada's 2015 restructuring introduced hurricane clauses into restructured debt, allowing debt service to be automatically re-profiled after a qualifying natural disaster. Official debt sustainability analysis later described this as a deliberate attempt to release resources rapidly when disaster financing needs were highest (International Monetary Fund & World Bank, 2019). After Hurricane Beryl in 2024, Grenada formally activated the clause, and the scheduled payments due in November 2024 and May 2025 were deferred and added to the bond's principal repayment profile (Government of Grenada, 2024; World Bank, 2025). The instrument did not erase losses, but it bought time.

Barbados offers a related lesson. During its 2018 to 2019 debt restructuring, it embedded natural disaster clauses into new debt instruments. IMF reporting at the time explicitly welcomed the move because it strengthened the country's protection against future weather-related shocks (International Monetary Fund, 2019a, 2019b). The point is important. These clauses are not merely for post-default restructurings. They can also be used prospectively to improve the shock-absorption features of new instruments where exposure is well understood.

Even so, contingency is not a cure for unsustainable debt. Cohen et al. (2020) are clear that state-contingent debt instruments are most useful when the underlying debt position is broadly viable and the main problem is uncertainty or shock timing. A clause cannot compensate for a debt stock that is already inconsistent with medium-term repayment capacity. In that situation, a deferral may simply postpone a restructuring. The instrument works best as part of a broader resilience framework, not as a substitute for fiscal repair or realistic debt treatment (Table 1).

One way to think about suitability is to ask whether the sovereign's problem is volatility or insolvency. State-contingent instruments are far more persuasive when debt is broadly sustainable and the concern is that a severe shock could force destructive adjustment at the wrong moment. They are much less useful when the debt stock is already inconsistent with medium-term repayment capacity. In that setting, a temporary payment deferral can postpone the recognition of a larger restructuring

Table 1 Trigger options for disaster and state-contingent clauses

Trigger type	Strength	Main weakness	Best use case
Parametric or model-based	Fast and objective once data are available	Basis risk if measured loss differs from actual fiscal stress	Catastrophe-linked instruments and insurance-style structures
Insurance payout-based	Verification is externally anchored	Depends on parallel insurance coverage and contract design	Hurricane or disaster clauses linked to CCRIF or similar facilities
Emergency declaration-based	Simple and administratively familiar	Greater discretion can raise investor concerns	Official-sector or closely monitored loan clauses
Macro trigger such as GDP or revenue shock	Can match broad fiscal stress	Slow data release and revision risk	Growth-linked or restructuring instruments rather than rapid disaster liquidity
Hybrid trigger	Balances objectivity and event relevance	More drafting and verification complexity	Sovereign bonds seeking faster relief with some guardrails
Fiscal-loss threshold	Closer link to budget stress	Measurement can be slow and contested	Official-sector or highly customized restructurings
Pre-defined payment cap	Limits uncertainty for investors	May reduce flexibility in extreme events	Bond clauses designed to protect pricing while preserving some relief
Investor disclosure package	Supports pricing and operational understanding	Weak communication can widen uncertainty	Repeat issuers seeking broader market familiarity with contingency

Source Author's own elaboration based on Cohen et al. (2020), Guerson (2021), Igan et al. (2021), and World Bank (2024)

problem without changing its substance. This is why both IMF discussion and operational guidance place these instruments inside a viability framework rather than treating them as a substitute for debt treatment where the numbers no longer add up (Cohen et al., 2020; Guerson, 2021; World Bank, 2021).

Another practical point is that clauses complement rather than replace other instruments. Countries still need cash buffers, disaster response planning, and in many cases insurance or catastrophe-risk transfer arrangements. State contingency fills a different gap by reducing near-term debt-service pressure on the liability side of the sovereign balance sheet. That is especially useful in small states and disaster-prone borrowers where one event can hit growth, revenue, and financing simultaneously. IMF work on climate-related financial instruments and the World Bank note on climate resilient debt clauses both support this layered view. The clause is one tool in a portfolio of resilience mechanisms, not a self-sufficient architecture (Ando et al., 2022; Belianska et al., 2022; World Bank, 2024).

Barbados and Grenada also illustrate a political economy advantage that is easy to miss. When the relief rule is known in advance, domestic policymakers can build disaster response planning around a defined financing assumption rather than hoping for exceptional creditor flexibility after the event. That changes behavior inside government as much as it changes expectations in markets. Finance ministries can estimate what a temporary suspension of debt service would actually free up, and line ministries can link that space to recovery priorities. The clause becomes more useful when it is operationalized before the shock arrives rather than celebrated only after activation becomes necessary (Government of Grenada, 2024; International Monetary Fund, 2019a, 2019b).

Designers should also think hard about frequency. A clause that can be activated repeatedly without clear limits may generate more pricing concern than expected, particularly where climate shocks are increasing in frequency. By contrast, a clause with a defined cap, a minimum time between activations, or a clearly stated maximum deferral period can be easier for investors to understand. The general principle is that rare but severe liquidity relief tends to be more credible than open-ended flexibility. Markets can price bounded contingencies more easily than ambiguous ones (Igan et al., 2021; World Bank, 2024).

3 Trigger Design: Clarity First, Elegance Second

Good trigger design begins by accepting that elegance is less important than verifiability. Policymakers are often tempted by triggers that appear economically comprehensive, but the market usually values clarity more highly than theoretical sophistication. A useful trigger should be externally observable, difficult to manipulate, quickly verifiable, and tightly aligned with the liquidity problem the clause is meant to solve. If any of those conditions is missing, the clause becomes harder to price and less likely to be trusted (Guerson, 2021).

Disaster clauses linked to third-party insurance determinations illustrate this principle well. Grenada's 2024 deferral claim relied on written confirmation tied to its catastrophe insurance arrangements, which reduced scope for interpretive disputes (Government of Grenada, 2024). The World Bank's Climate Resilient Debt Clause follows a similar logic in official lending. Its updated 2024 product note explains that eligible borrowers may defer principal, interest, and certain other charges for up to two years following a qualifying event, with defined eligibility and activation conditions. That is a narrow clause, but its narrowness is precisely what makes it legible (World Bank, 2024).

By contrast, broad macro triggers can be informative without being operationally timely. Output or revenue collapses may capture the overall fiscal impact of a shock, yet they tend to be observed with delay and are vulnerable to revision. That makes them less suitable for immediate post-disaster liquidity relief. They may still have a place in state-contingent instruments designed for restructurings or growth-sharing arrangements, but the design problem is different there. Speed matters more in a disaster clause than comprehensiveness (Cohen et al., 2020; Guerson, 2021).

Another practical rule is to align the trigger with the intended coverage. A clause that covers one payment cycle can tolerate a narrow, high-confidence trigger. A clause that affects multiple payment dates or a large share of the debt portfolio demands stronger safeguards and clearer creditor coordination. Some recent proposals in international policy circles push for broader "pause clauses" in sovereign bonds. The direction is interesting, but the mature lesson from actual use remains conservative: it is better to have a clause that triggers predictably in a severe event than one that promises flexibility but becomes contested at the very moment it is needed most (International Monetary Fund & World Bank, 2019; World Bank, 2024).

Trigger design is really a trade-off between basis risk and operational certainty. A declaration-based trigger tied to a recognized disaster event can be faster to verify, but it may be too crude if the fiscal impact varies widely across events. A loss-based trigger can in principle align more closely with actual damage, yet it may take longer to measure and may create room for dispute. Parametric structures improve speed and objectivity, but they can misfire when the chosen indicator does not reflect the actual fiscal burden. This is why good design is usually less about theoretical precision than about credible verification, timeliness, and legal simplicity. If the trigger is too difficult to observe, the liquidity value of the clause is weakened at exactly the wrong moment (Guerson, 2021; Igan et al., 2021; World Bank, 2024).

The contract also has to define the relief mechanics in painful detail. How long are payments deferred. Are deferred coupons capitalized or merely pushed out. Is there an annual cap on activations. Can the clause be triggered multiple times during the life of the bond. What notice must the sovereign provide. How quickly must the trustee or paying agent act. These are not technical footnotes. They determine whether the clause behaves as a predictable cash-flow tool or becomes a source of operational confusion under stress. Debt-management guidance and recent practice both suggest that markets respond better when activation mechanics are narrow, explicit, and embedded in standard documentation rather than negotiated as bespoke side arrangements (Cohen et al., 2020; World Bank and International Monetary Fund, 2019; World Bank, 2024).

One practical calibration issue is the relationship between trigger severity and the amount of relief. A severe event may justify a longer deferral or broader inclusion of principal and interest, while a moderate event may justify only coupon deferral. Building that proportionality into the documentation can make the instrument more usable, but only if it does not introduce excessive complexity. Every layer of sophistication has to earn its keep by improving clarity or fit. Otherwise, what looks like customization can turn into execution risk under stress (Cohen et al., 2020; Guerson, 2021).

Data source choice matters for the same reason. Triggers tied to independent, widely recognized data or institutions are easier to trust. Where governments propose relying on domestic declarations or damage estimates alone, they may face more investor skepticism unless the verification path is tightly specified. The attraction of objectivity is not mainly legal elegance. It is speed. The whole point of the clause is to preserve liquidity quickly after a shock. A verification process that drags on for months weakens the economic rationale even if the trigger is theoretically sound (Belianska et al., 2022; World Bank, 2024).

4 Pricing and Documentation: Where Credibility Is Tested

Investors rarely oppose resilience in principle. Their concern is whether contingency changes expected value, increases uncertainty, or creates legal ambiguity. The pricing literature suggests that premia on state-contingent sovereign instruments need not be prohibitive when the clause reduces default risk and is clearly specified, but the result depends on probability, severity, transparency, and the relationship between the contingent instrument and the rest of the debt stock (Igan et al., 2021). A clause that is easy to understand and infrequently triggered may therefore be cheaper than policymakers initially assume.

Documentation does much of the real work. The contract needs a precise event definition, a named verification process, a notice period, a calculation rule for deferred amounts, and a statement on how maturity extension or capitalization of interest will operate. It also needs to clarify the interaction with other contractual terms such as collective action clauses, pari passu commitments, and events of default. Poorly coordinated documentation can recreate the uncertainty the clause was supposed to reduce (Cohen et al., 2020).

Portfolio coverage is another neglected issue. A sovereign that inserts a resilience clause into one loan but leaves the rest of its portfolio rigid may still face the same liquidity strain after a shock. The clause helps, but the balance-sheet effect remains partial. This is why the World Bank has stressed scalability in its recent work on climate-resilient debt clauses and why Caribbean experience is often discussed alongside Cat DDOs, insurance, and emergency liquidity planning rather than as a stand-alone solution (World Bank, 2024, 2025).

Transparency before issuance is just as important as enforcement after the event. Countries need to explain why the clause is being used, what risks it is meant to address, how it interacts with overall debt sustainability, and how any deferred debt service will be treated in official debt statistics and cash management. A resilience feature becomes easier for markets to absorb when it is presented as part of a disciplined debt strategy rather than as a last-minute appeal to exceptionalism (Igan et al., 2021; Knight & Northfield, 2020).

Pricing remains the hardest issue because markets have only a limited history from which to infer premia. Investors are giving up some cash-flow certainty, so the protection is unlikely to be free. Yet the premium is not likely to be constant across issuers or structures either. It should depend on the probability of activation, the transparency of the trigger, and the degree to which the affected investor base can diversify the relevant shock. That is why theoretical pricing work and IMF analysis emphasize issuer-specific context rather than any universal number. A highly exposed borrower with a well-designed, narrow clause may still be better off paying some premium if the clause materially reduces the risk of a disorderly post-disaster financing squeeze (Ando et al., 2022; Guerson, 2021; Igan et al., 2021).

Documentation quality therefore carries financial value. Clear notice provisions, independent verification, explicit treatment of deferred payments, and careful drafting around cross-default and pari passu interactions reduce uncertainty for both sides. Grenada's recent experience is instructive because the activation process was not merely a theoretical possibility. It required notice, verification, and operational execution within an already stressed environment. That real-world activation reminds policymakers that the clause is only as useful as the surrounding legal and settlement infrastructure. Good drafting is not window dressing. It is the difference between fast contractual relief and a new round of interpretation disputes when time is most scarce (Government of Grenada, 2024; World Bank, 2024; World Bank and International Monetary Fund, 2014).

Investor communication should therefore be treated as part of documentation, not as a separate public-relations exercise. Borrowers need to explain why the clause exists, how often it could plausibly be triggered, what the maximum payment effect would be, and how it fits within the broader borrowing strategy. If those questions are left unanswered, investors may fill the gap with worst-case assumptions. Sovereign investor-relations practice already points in this direction. State contingency raises the need for disciplined explanation because uncertainty about the rule can be almost as costly as the rule itself (Igan et al., 2021; Knight and Northfield, 2020).

A related caution is that clauses should not be used to mask weak fiscal preparation. If a government is routinely exposed to natural disasters but has done little to improve budget flexibility, reserve policy, or cash management, the clause may end up compensating for broader shortcomings. The more persuasive use case is where the sovereign has already built other resilience tools and uses contractual contingency to cover the specific debt-service timing problem that remains. That layered use is what makes the clause look like prudence rather than desperation (World Bank, 2021; World Bank and International Monetary Fund, 2019).

5 When Clauses Add Value and When They Do Not

State-contingent clauses are most valuable where four conditions hold. First, the country faces recurrent, externally observable shocks such as hurricanes, floods, or severe droughts. Second, liquidity pressures are likely to spike quickly after the event. Third, the debt position is broadly serviceable outside the disaster state. Fourth, the issuer can support the clause with strong disclosure and a wider disaster-risk financing architecture. Small island states and climate-exposed lower-middle-income countries often fit this profile more closely than larger diversified economies (Ando et al., 2022; World Bank, 2024).

Their value is lower when debt is already structurally unsustainable, when the trigger is too subjective, or when the clause covers only a trivial share of the portfolio. It also falls when institutions cannot produce timely notices, maintain creditor communication, or align the clause with budget execution and liquidity management. In those settings, the political appeal of the instrument can outrun its operational usefulness. That is one reason why official-sector clauses and restructuring-linked clauses have moved faster than broad-based market standardization.

A sensible adoption path is incremental. Official loans and MDB-supported instruments provide a controlled environment in which governments can build contractual familiarity, operational routines, and reporting capacity. Private market adoption can follow where exposure, market access, and investor dialogue support it. The recent World Bank productization of CRDCs is important precisely because it lowers the implementation barrier and turns an abstract resilience principle into a standard contractual feature for eligible borrowers (World Bank, 2024).

The broader lesson is modest but powerful. Disaster and state-contingent clauses do not make debt painless, and they do not replace fiscal space. What they can do is reduce the odds that a temporary climate shock is handled through economically damaging improvisation. For highly exposed sovereigns, that is already a meaningful advance. The design challenge is to keep the clauses narrow enough to be credible and broad enough to matter once triggered.

Where, then, do these clauses add the most value. The answer is usually in highly exposed sovereigns that are market borrowers, have debt positions that are still broadly manageable, and want to reduce the fiscal damage of exogenous shocks without reopening a restructuring every time a severe event occurs. Barbados and Grenada remain the most visible reference points because they show both the strategic rationale and the operational demands involved. In each case, the clause made sense because climate and disaster exposure was material enough to threaten liquidity, yet the sovereign still needed an instrument that could coexist with a broader financing and debt strategy (International Monetary Fund, 2019a, 2019b; International Monetary Fund and World Bank, 2019; World Bank, 2025).

The instrument is much less persuasive when any of the underlying conditions are missing. If debt is already unsustainable, the clause may only delay recognition of a larger restructuring need. If the trigger is too broad, pricing and credibility become harder. If documentation is vague, the relief can prove slower than expected. That is why the design principle is deceptively simple. The clause should be narrow enough to be trusted, specific enough to be executed quickly, and integrated enough to fit the sovereign's existing debt-management and resilience framework. Anything looser risks turning a promising resilience tool into a source of fresh uncertainty (Cohen et al., 2020; Guerson, 2021; World Bank, 2024).

Standardization may prove to be one of the biggest medium-term gains. As documentation becomes more familiar, investors and issuers should spend less time debating basic mechanics and more time calibrating which clause fits which exposure profile. That kind of standard-setting would help move the instrument from novelty toward normal sovereign risk management.

6 Conclusion

Disaster and state-contingent clauses are no longer a theoretical curiosity. Caribbean experience, IMF research, and recent World Bank operational work now provide enough evidence to say what these clauses are good for and what they are not. They are liquidity instruments that can create breathing space after a severe exogenous shock. They are not substitutes for solvency treatment, fiscal adjustment, or ordinary debt management discipline.

Their effectiveness depends less on rhetorical ambition than on contract design. Clear triggers, transparent documentation, realistic portfolio coverage, and strong institutional follow-through determine whether a clause will be priced, trusted, and used well. The strongest sovereign cases are therefore likely to be those that combine the clause with insurance, contingent credit, cash planning, and credible debt strategy. In that setting, contingency becomes a disciplined resilience tool rather than a slogan.

The policy implication is that governments should prepare these clauses before they need them. That means embedding the possibility of state contingency in annual borrowing plans, model legal documentation, and investor communication rather than treating it as an emergency innovation. Operational readiness matters just as much as contract wording. Debt managers, legal advisers, and paying agents should know in advance how evidence will be assembled, how notice will be issued, and how revised debt-service schedules will be communicated to investors. Preparing that machinery during calm periods is part of what makes the clause credible in the first place (World Bank and International Monetary Fund, 2014, 2019; World Bank, 2021).

Seen this way, disaster and state-contingent clauses are not exotic accessories. They are one layer in a mature sovereign resilience toolkit that also includes prudent borrowing strategy, cash buffers, insurance, and climate-sensitive fiscal planning. Their value lies in disciplined design and disciplined use. Done well, they can preserve fiscal space when it is most scarce. Done badly, they can add legal complexity without delivering reliable relief.

There is also a useful comparison with catastrophe bonds and insurance-style instruments. Those tools transfer risk to different counterparties and can deliver rapid payouts, but they do not directly alter the sovereign's debt-service schedule.

State-contingent clauses solve a narrower problem on the liability side. They are often most valuable when paired with other instruments rather than chosen instead of them. The toolkit works better when each layer has a distinct job.

For that reason, the strongest chapters in sovereign finance are no longer about whether one instrument will solve resilience on its own. They are about how different instruments can be combined without creating fresh opacity. Disaster clauses have earned a place in that discussion because they respond to a real financing gap. Their value, however, remains inseparable from design discipline.

The strongest sovereigns will probably be those that can explain these clauses not as requests for leniency, but as disciplined arrangements for handling objectively defined shocks. That framing turns contingency into part of prudent borrowing rather than an exception to it.

In that sense, documentation quality is a policy choice as much as a legal one. Clear clauses tell the market that the sovereign has thought through how it will behave under stress before that stress arrives.

Over time, the real achievement may be cultural as much as contractual. Once issuers and investors become familiar with disciplined contingency, relief after a major shock can be handled more as agreed procedure and less as an ad hoc negotiation.

References

Ando, S., Roch, F., Wiriadinata, U., & Fu, C. (2022). *Sovereign climate debt instruments: An overview of the green and catastrophe bond markets* (IMF Staff Climate Notes, 2022(004)). https://doi.org/10.5089/9798400210006.066

Belianska, A., Bohme, N., Cai, K., Diallo, Y., Jain, S., Melina, G., Mitra, P., Poplawski-Ribeiro, M., & Zerbo, S. (2022). *Climate change and select financial instruments: An overview of opportunities and challenges for Sub-Saharan Africa* (IMF Staff Climate Notes, 2022(009)). https://doi.org/10.5089/9798400225208.066

Cohen, C., Anthony, M., Best, T., Breuer, P., Miao, H., Myrvoda, A., & Togo, E. (2020). *The role of state-contingent debt instruments in sovereign debt restructurings* (IMF Staff Discussion Notes, 2020(006)). https://doi.org/10.5089/9781513556482.006

Government of Grenada. (2024). *Notice to bondholders: Deferral claim.* Ministry of Finance.

Guerson, A. D. (2021). *Optimal state contingent sovereign debt instruments* (IMF Working Papers, 2021(230)). https://doi.org/10.5089/9781513595917.001

Igan, D. O., Kim, T., & Levy, A. (2021). *The premia on state-contingent sovereign debt instruments* (IMF Working Papers, 2021(282)). https://doi.org/10.5089/9781616357009.001

International Monetary Fund. (2019a). *Barbados: 2019 Article IV consultation, second review under the extended arrangement under the extended fund facility, request for modification of performance criteria, and financing assurances review* (IMF Country Report No. 19/370).

International Monetary Fund. (2019b, December 16). *IMF executive board concludes 2019 article IV consultation with barbados and completes second review under the EFF arrangement* (IMF Press Release No. 19/462).

International Monetary Fund, & World Bank. (2019). *Grenada joint World Bank–IMF debt sustainability analysis.* International Monetary Fund and World Bank.

Knight, J., & Northfield, B. (2020). *Sovereign investor relations: From principles to practice* (IMF Working Papers, 2020(204)). https://doi.org/10.5089/9781513556369.001

World Bank. (2021). *Debt management performance assessment methodology: 2021 edition*. World Bank.

World Bank. (2024). *Climate resilient debt clause (CRDC) note*. World Bank Treasury.

World Bank. (2025). *Grenada climate resilient debt clause activation and recovery financing overview*. World Bank.

World Bank, & International Monetary Fund. (2014). *Revised guidelines for public debt management*. World Bank.

World Bank, & International Monetary Fund. (2019). *Developing a medium-term debt management strategy (MTDS): Updated guidance note for country authorities*. World Bank and International Monetary Fund.

Liability Management, Swaps, and Guarantees Climate Risk and Sovereign Green Bonds

Miriam R. Aziz

1 Introduction

Climate risk has become part of sovereign liability management, not merely a background condition for environmental policy. When floods, heat stress, transition costs, or delayed adaptation affect output and public balances, debt managers face a harder task: they must fund climate investment while preserving market access and avoiding a rise in refinancing risk. Recent evidence points in that direction. Climate exposure can be associated with higher sovereign credit risk, and transition capacity can shape the cost of borrowing through institutional and fiscal channels (Arena et al., 2026; Pan et al., 2026; Seghini, 2026). In this setting, the question is not whether climate policy belongs in debt strategy. It already does. The more difficult issue is how governments can design liabilities that are affordable, credible, and resilient under climate uncertainty.

Green bonds entered this debate as a visible instrument, but they should not be treated as a complete solution. Their appeal is clear: proceeds are earmarked for environmental projects, investor demand can widen, and some issuers may obtain a modest greenium. Yet the pricing advantage is not automatic. Ando et al. (2024) find a sovereign greenium that differs across advanced and emerging markets, while Chesini (2024) shows that European sovereign green bonds can support the net-zero transition only when issuance is backed by credible policy and project governance. Evidence from emerging economies reaches a similar conclusion in a different context: sovereign green bonds can mobilize capital for renewable energy, clean transport, water, and ecosystem projects, but they may also add to the public debt burden if social equity and development priorities are weakly embedded (Owusu-Mante et al., 2025).

M. R. Aziz (✉)
Department of Management, INSEEC Grande Ecole, Paris, France
e-mail: maziz@inseec.com

H. Jarrar et al. (eds.), *Sovereign Sustainable Finance*, Sustainable Finance,
https://doi.org/10.1007/978-3-032-33239-4_6

This chapter links that evidence to the title themes of liability management, swaps, and guarantees. Liability management concerns the structure of debt, including maturity, currency, coupon profile, refinancing calendar, and investor composition. Swaps help adjust that structure after issuance by managing interest-rate or currency exposures. Guarantees, in turn, can reduce perceived credit risk at the project or instrument level, especially when private investors need assurance that climate projects are bankable. The discussion therefore treats sustainable debt as part of a broader financing architecture, rather than as a separate ESG niche. This framing is consistent with research showing that sustainable debt markets now include green, social, sustainability, and sustainability-linked instruments, each priced through a combination of label effects, liquidity, risk, and issuer credibility (Becker et al., 2025; Poggensee, 2025; Vulturius et al., 2024).

The discussion also treats these instruments as linked choices rather than as separate policy tools. A debt office may issue a green bond to fund a climate project, use a swap to adjust the interest-rate or currency profile of that liability, and rely on a guarantee to improve the bankability of the underlying investment. Those steps are technical, but the policy logic is simple: climate finance must not create a liability structure that becomes harder to service under stress. This is why the chapter focuses on the fit between instrument design and fiscal capacity. Sustainable debt is useful only when the repayment profile, the project pipeline, and the risk-transfer tools point in the same direction.

2 Climate Risk as a Sovereign Liability Problem

A government's liabilities are exposed to climate risk through several channels at once. Physical risk can damage infrastructure, lower tax revenue, and increase emergency spending. Transition risk can raise investment needs before the fiscal benefits of decarbonization are visible. That timing mismatch is important for debt managers because markets price the near-term capacity to service debt, not only the long-term social return of climate action. Arena et al. (2026) show that investors reward credible institutional and economic readiness with lower borrowing costs, but they may view some social adaptation obligations as fiscal constraints. Pan et al. (2026) also find that climate exposure is linked to sovereign credit risk through economic and fiscal channels, including GDP per capita and government debt dynamics.

The timing problem is especially visible when climate projects require large upfront spending but generate benefits slowly. Adaptation infrastructure, grid upgrades, resilient water systems, and clean transport may reduce future losses, yet their fiscal payoff is rarely immediate. Investors may therefore price the debt before the risk reduction becomes observable. Arena et al. (2026) describe this problem as a fiscal channel: climate vulnerability can raise borrowing costs, and higher borrowing costs can then reduce the government's capacity to finance the very investments that would lower vulnerability. For liability managers, the implication is concrete.

Maturity choice, redemption concentration, and coupon structure should be designed around this lag rather than around annual funding targets alone.

Credit ratings make the picture less settled. Andersen (2025) finds no systematic evidence that major sovereign ratings have responded to long-horizon climate risk, which suggests that rating agencies may still prioritize short- to medium-term fundamentals over distant climate scenarios. Other studies point toward a stronger connection between climate risks, sovereign ratings, and long-term borrowing costs, particularly when transition and physical exposures are assessed more directly (De Angelis et al., 2026; Erhart et al., 2026). The tension is useful rather than contradictory. It suggests that climate risk is entering sovereign finance unevenly: investors, rating agencies, and policymakers may recognize similar risks but convert them into prices, ratings, and debt strategies at different speeds.

A second implication concerns the boundary between market pricing and public accounting. Climate-related liabilities may appear through explicit debt, guaranteed obligations, disaster reconstruction costs, or weaker tax revenues after shocks. These items do not all enter the balance sheet in the same way, but investors may eventually connect them when evaluating solvency. Andersen (2025) shows that rating agencies may not yet respond systematically to long-horizon climate risks, while other work finds stronger pricing or rating effects when climate exposure is modeled more directly (De Angelis et al., 2026; Erhart et al., 2026). That gap makes conservative liability management more important, not less, because delayed recognition can turn into abrupt repricing.

For liability management, this uneven pricing creates both a risk and an opportunity. If markets underprice climate risk today, sovereigns may delay adaptation and face sharper repricing later. If markets overreact to weak climate indicators, vulnerable countries may pay higher spreads before they can finance the investments needed to reduce vulnerability. That problem is sharper for emerging markets, where fiscal space is narrower and climate investment needs are large (Nevatia, 2026; Romagnoli & Santini, 2025). Subnational governments face a related difficulty because mitigation and adaptation investments are often implemented locally, while financing capacity and guarantees may depend on national frameworks (de Mello & Ter-Minassian, 2024). Carbon-market revenues and emissions-trading schemes can support fiscal planning, but they do not replace disciplined debt management or credible public investment selection (Kukah, 2025).

3 Green Bonds, Swaps, and the Cost of Public Debt

Sovereign green bonds can alter the composition of public liabilities by attracting investors who are willing to hold climate-aligned debt, sometimes at lower yields. The evidence, however, does not support a simple claim that green debt is always cheaper. Ando et al. (2024) report a sovereign greenium, with stronger effects in emerging markets, while García-Escobar et al. (2026) find that EU-issued ESG bond yield differences are better explained by liquidity than by the ESG label itself. Di Tommaso et al. (2025) likewise question whether governments are always greener than other issuers, which matters because sovereign labels can carry political credibility but also larger risks of weak additionality. Put differently, greenium is a market outcome, not a legal property of the bond.

The greenium should also be read against issuance mechanics. A low yield can reflect strong ESG demand, but it can also reflect currency choice, benchmark scarcity, bond size, or the liquidity conditions of the secondary market. Chesini (2024) finds that European sovereign green bonds tend to show a small greenium and that euro-denominated issuance can be associated with lower yields; however, the same study stresses that green bond markets face hurdles linked to standards, greenwashing concerns, and the sovereign principle of fiscal fungibility. Ando et al. (2024) reach a related but more quantified result, estimating a modest average greenium in advanced economies and a larger one in emerging markets. The shared message is cautious: pricing benefits exist, but they require institutional support.

Swaps fit here because green debt does not remove the financial risks attached to public borrowing. A sovereign may issue a long-maturity green bond to fund adaptation infrastructure, but its debt-service profile still depends on interest rates, currency denomination, and investor demand at future rollover dates. In practice, interest-rate swaps can convert fixed-rate exposure into floating-rate exposure, or the reverse, when the debt office needs to rebalance duration. Currency swaps can reduce exchange-rate risk if international ESG demand pushes an issuer toward foreign-currency bonds. Climate-linked or sustainability-linked structures may also create contingent cash-flow questions, although current evidence on sustainability-linked bonds remains mixed (Anderson & Kish, 2024; Oude Lansink & Kapelko, 2025; Poggensee, 2025). The point is not that swaps make green bonds safer by themselves; they give debt managers a way to align the liability profile with the fiscal risks of climate projects.

Used carefully, swaps can prevent a green issuance program from becoming too dependent on one investor segment or one currency. Suppose foreign ESG investors prefer a dollar or euro bond while the sovereign's revenues are mostly domestic. The green label may attract demand, but the currency mismatch remains. A currency swap can reduce that mismatch; an interest-rate swap can rebalance duration if market conditions change after issuance. Yet this is not costless risk removal. Swap counterparties, collateral terms, rollover dates, and mark-to-market movements create their own governance demands. The debt office therefore needs a clear rule: hedging should reduce identifiable balance-sheet risk, not become a separate source of speculative exposure.

Firm-level evidence adds a useful warning. Guesmi et al. (2025) find that climate-exposed firms are more likely to issue green bonds, but those bonds may carry higher coupon rates and longer maturities, and issuance is associated with improved ESG scores rather than lower emissions. Sovereigns face an analogous credibility test. If proceeds finance projects that look green on paper but do not reduce vulnerability or emissions, the liability remains while the climate benefit is weak. Country-level work on scarcity effects and climate risk in green bonds reinforces that the investor base is sensitive to both climate exposure and instrument supply (Romagnoli & Santini,

2025). That said, a carefully structured green bond program can still support liability management by smoothing the maturity profile, enlarging demand, and creating benchmarks for domestic sustainable debt markets.

India's experience illustrates the gap between policy intent and immediate market pricing. The Reserve Bank of India's sovereign green bond trading facilitation through the International Financial Services Center was designed to widen foreign participation, improve liquidity, and strengthen price discovery. Mahfooz and Singh (2026) find that the short-term market reaction was largely neutral to mildly negative, suggesting that infrastructure reforms do not instantly translate into a valuation gain. The lesson is narrow but important: market access mechanisms, swap capacity, transparent settlement systems, and investor outreach must operate together. A green bond that is hard to trade, poorly reported, or exposed to unmanaged currency risk will struggle to deliver a durable funding advantage.

4 Guarantees, Credibility, and Market Spillovers

Guarantees address a different weakness in climate finance: the gap between project need and investor risk appetite. Many mitigation and adaptation projects produce public benefits over long horizons, while investors require predictable cash flows and enforceable protections. Guarantees can improve the credit profile of selected projects, reduce perceived default risk, and crowd in private capital, particularly where the sovereign balance sheet alone cannot carry the full transition cost. Tyson and Weiss (2025) emphasize that net-zero investment requires public–private collaboration, policy support, and financial innovation because public budgets cannot fund the transition on their own. In this sense, guarantees are not a substitute for fiscal discipline; they are a way to allocate risk more deliberately.

Still, guarantees can create hidden liabilities if governments use them without careful monitoring. A guarantee may not appear as direct debt at issuance, but it can become a fiscal cost when a project underperforms or macroeconomic conditions deteriorate. That is why liability management must account for contingent liabilities alongside conventional bonds, green bonds, and swaps. Emerging-economy evidence is especially relevant here. Owusu-Mante et al. (2025) show that sovereign green bonds can mobilize private climate capital, but debt burden and equity concerns remain central. Nevatia (2026) frames sovereign green bonds in emerging markets as risk-mitigating sustainability instruments, yet that risk-mitigation role depends on whether proceeds are allocated to projects that strengthen long-term repayment capacity rather than only satisfy reporting categories.

The guarantee question is equally sensitive because climate projects often sit between public value and commercial uncertainty. A sovereign guarantee can make a renewable-energy project, a clean-transport concession, or a resilient-water investment more acceptable to private investors. That can be efficient when the government is best placed to absorb policy or demand risk. The danger appears when guarantees are granted to accelerate issuance without rigorous project appraisal. Then the state may appear to mobilize private capital while quietly accumulating contingent liabilities. Owusu-Mante et al. (2025) make this concern visible in emerging economies: sovereign green bonds can mobilize capital, but their development value depends on equity, affordability, and alignment with national priorities.

Market spillovers complicate the design of guarantees and hedges. Green finance assets are not insulated from stress in broader markets; tail-risk transmission can increase when volatility rises (Naeem et al., 2024). Cross-regional evidence also shows that green bonds may offer hedging or diversification benefits, but the strength of those benefits varies by region and market condition (Karkowska & Urjasz, 2025). Related work on green stocks, green bonds, and policy uncertainty reaches a similar conclusion: sustainable assets can be connected to uncertainty shocks rather than immune from them (Syed, 2025). Pedini (2025) adds that ESG investments may act as hedges, diversifiers, or safe havens only in specific conditions. For sovereign debt offices, this means guarantees and swaps should be stress-tested, not assumed to work in every market state.

There is a portfolio dimension as well. Green bonds, swaps, and guarantees should be reviewed together because each one changes the government's exposure under stress. A guaranteed project may fail when revenues weaken; a swap may require collateral when rates or exchange rates move sharply; a green bond may become expensive to refinance if climate credibility deteriorates. Naeem et al. (2024) and Karkowska and Urjasz (2025) both point to the importance of stress and spillover effects in green finance markets. For that reason, a narrow focus on issuance-day pricing is insufficient. The better test is whether the full liability package remains manageable under adverse climate, market, and fiscal scenarios.

There is also a signaling effect beyond the sovereign market. Sustainable sovereign issuance can shape corporate sustainable bond markets by providing benchmarks, liquidity references, and policy signals (Kim et al., 2026). In Latin America, ESG bond evidence suggests that sustainability labels can produce value effects, but investor response depends on credibility and market context (Arévalo et al., 2024). The wider debt-capital-market literature points in the same direction: labels matter most when they are supported by transparent use of proceeds, credible issuer behavior, and liquid secondary markets (Becker et al., 2025). Guarantees can strengthen this ecosystem when they make climate projects investable; they can weaken it when they mask risk or encourage issuance without real climate performance.

5 Governance, Taxonomy Alignment, and Research Gaps

Governance is the thread that connects liability management, swaps, and guarantees. A government may issue a green bond, hedge the currency exposure, and guarantee a project, but investors will still ask whether the framework is credible. Taxonomy alignment is one answer, although it brings its own difficulties. Misalignment across sustainable finance taxonomies can disrupt cross-border capital flows because investors face uncertainty over what qualifies as green or sustainable (Chan et al., 2025). Green bond market practices also reflect a moral balance between environmental purpose and financial value, which becomes fragile when reporting is vague or proceeds are difficult to trace (Bracking, 2024).

Reporting regimes are not administrative details. They determine whether investors can distinguish a credible climate-finance strategy from a well-labeled borrowing program. The EU evidence is useful here because it shows that ESG labels do not automatically dominate liquidity and volatility factors in pricing (García-Escobar et al., 2026). Misaligned taxonomies can also reduce cross-border capital flows by making eligibility uncertain (Chan et al., 2025). In practice, a government that wants lower borrowing costs must show more than eligible project categories. It must show how proceeds are selected, how outcomes are measured, how unspent balances are handled, and how climate claims are reconciled with the broader fiscal framework.

Institutional credibility shapes the same issue from the investor side. Asset managers' net-zero target-setting depends on norms, incentives, and market pressures, not only on climate science (Buentjen et al., 2025). The international political economy of green finance therefore matters because standards, disclosure rules, and investment mandates influence where capital moves and how risk is interpreted (Sharma & Babic, 2025). ESG has also become more deeply embedded in credit assessment as rating agencies and data providers expand their sustainable-finance capacity, including through acquisitions and methodological changes (Malone, 2025). For sovereigns, the implication is direct: debt instruments gain credibility when they are tied to institutions that can measure, report, and enforce climate commitments.

The research agenda remains open. A systematic review of green debt finds expanding work on pricing, issuance, market development, and environmental integrity, but it also identifies the need for stronger links between financial outcomes and real-economy effects (Flottmann et al., 2025). That gap is visible in the greenium debate. Some studies find pricing benefits, others attribute yield differences to liquidity, and still others warn that labels may improve ESG profiles without proving emissions reductions (Ando et al., 2024; García-Escobar et al., 2026; Guesmi et al., 2025). Future research should therefore connect liability-level decisions—maturity, currency, coupon, swaps, guarantees, and contingent liabilities—with climate outcomes that can be independently verified.

This point matters for the future research agenda because the next question is no longer simply whether green bonds are cheaper. A more relevant question is whether climate-linked liability strategies reduce sovereign fragility over time. That requires evidence on refinancing risk, currency exposure, contingent guarantees, and post-issuance project outcomes. Flottmann et al. (2025) call for stronger links between green debt research and real-economy effects; this chapter extends that call to debt management. The strongest empirical work will likely compare instruments not only by yield, but also by their contribution to fiscal resilience, climate adaptation capacity, and investor trust after market stress.

6 Conclusion

The chapter's main claim is that climate finance belongs inside sovereign liability management. Climate risk can affect creditworthiness, fiscal balances, ratings, spreads, and investor behavior, even if markets do not price every channel consistently. Green bonds offer a useful instrument because they can mobilize dedicated capital, signal policy intent, and create sustainable-finance benchmarks. Their limits are just as important. A green label does not eliminate interest-rate risk, currency exposure, refinancing pressure, weak liquidity, or doubts about environmental additionality.

Swaps and guarantees complete the liability-management logic. Swaps allow debt offices to adjust duration and currency exposure after issuance, which is valuable when green bonds are issued in maturities or currencies shaped by investor demand rather than by the government's ideal risk profile. Guarantees can improve project bankability and crowd in private capital, but they must be treated as contingent liabilities and stress-tested under adverse scenarios. Used carefully, these instruments help connect climate ambition with fiscal resilience. Used loosely, they can hide risk and weaken trust.

A stronger framework would therefore combine four elements: credible climate policy, transparent green-bond reporting, disciplined hedging of financial exposures, and prudent management of guarantees. The literature reviewed here does not suggest that sustainable debt is always cheaper or safer. It suggests something more useful for policymakers: sustainable debt can become cheaper, safer, and more effective when its design is integrated with liability management and when environmental claims are backed by measurable outcomes.

For policymakers, the practical conclusion is not to issue more labeled debt for its own sake. The better objective is to build a liability structure that can finance climate investment without weakening solvency. That means using green bonds where they widen demand or support credible project pipelines, swaps where they reduce measurable financial exposures, and guarantees where they unlock projects that improve long-term resilience. The discipline lies in treating each instrument as part of the same balance sheet. When that discipline is absent, the public sector can look greener while becoming more financially fragile. When it is present, sustainable finance can support both climate policy and debt sustainability.

References

Andersen, T. B. (2025). Sovereign debt in a warming world: Are credit ratings responding to climate risks? *Kyklos, 78*(4), 1479–1495. https://doi.org/10.1111/kykl.70002

Anderson, A.-M., & Kish, R. J. (2024). Rewarding performance through sustainability-linked bonds. *Economic Affairs, 44*(2), 294–319. https://doi.org/10.1111/ecaf.12636

Ando, S., Fu, C., Roch, F., & Wiriadinata, U. (2024). How large is the sovereign greenium? *Oxford Bulletin of Economics and Statistics, 86*(6), 1472–1483. https://doi.org/10.1111/obes.12619

Arena, C., Basiglio, S., & Comandé, A. (2026). The impact of climate risk on sovereign debt: Evidence from OECD economies. *Journal of Environmental Management, 402*, Article 128851. https://doi.org/10.1016/j.jenvman.2026.128851

Arévalo, G., González, M., Guzmán, A., & Trujillo, M.-A. (2024). The value effect of sustainability: Evidence from Latin American ESG bond market. *Journal of Sustainable Finance & Investment, 14*(3), 516–537. https://doi.org/10.1080/20430795.2024.2344527

Becker, A., Fatica, S., & Rancan, M. (2025). Not only green: Sustainability and debt capital markets. *Journal of International Money and Finance, 154*, Article 103319. https://doi.org/10.1016/j.jimonfin.2025.103319

Bracking, S. (2024). Green bond market practices: Exploring the moral balance of environmental and financial values. *Journal of Cultural Economy, 17*(3), 279–296. https://doi.org/10.1080/17530350.2024.2312864

Buentjen, C., Perkins, R., & Sullivan, R. (2025). Net-zero norms in sustainable finance: What explains asset managers' target-setting? *Journal of Sustainable Finance & Investment, 15*(4), 954–985. https://doi.org/10.1080/20430795.2025.2520524

Chan, K. J. D., Mok, L. W., & Lau, P. C. C. (2025). Leakage in the common ground: How misalignment in sustainable finance taxonomies impacts cross-border capital flows. *Journal of Sustainable Finance & Investment, 15*(3), 688–713. https://doi.org/10.1080/20430795.2025.2481954

Chesini, G. (2024). Can sovereign green bonds accelerate the transition to net-zero greenhouse gas emissions? *International Advances in Economic Research, 30*(2), 177–197. https://doi.org/10.1007/s11294-024-09900-6

de Mello, L., & Ter-Minassian, T. (2024). Subnational investments in mitigation and adaptation to climate change: Some financing and governance issues. *Public Finance and Management, 23*(3), 113–131. https://doi.org/10.1177/15239721241259570

De Angelis, L., Monasterolo, I., & Zanin, L. (2026). Look up and ahead: How climate scenarios affect European sovereign credit risk. *International Review of Economics & Finance, 107*, Article 105057. https://doi.org/10.1016/j.iref.2026.105057

Di Tommaso, C., Perdichizzi, S., Vigne, S., & Zaghini, A. (2025). Is the Government always greener? *Energy Economics, 152*, Article 108976. https://doi.org/10.1016/j.eneco.2025.108976

Erhart, S., Bethlendi, A., & Erhart, K. (2026). The impact of climate risks on European sovereign credit ratings and long-term borrowing costs. *International Review of Economics & Finance, 106*, Article 105021. https://doi.org/10.1016/j.iref.2026.105021

Flottmann, C., Köchling, G., Neukirchen, D., & Posch, P. N. (2025). *Green debt: A systematic literature review and future research agenda*. Advance online publication. https://doi.org/10.1007/s11301-025-00511-x

García-Escobar, J., Fernández-Guadaño, J., & Mascareñas, J. (2026). Empirical comparative analysis of the performance of European Union conventional and ESG bonds. *Discover Sustainability, 7*, Article 204. https://doi.org/10.1007/s43621-025-02437-6

Guesmi, K., Makrychoriti, P., & Pyrgiotakis, E. G. (2025). Climate change exposure and green bonds issuance. *Journal of International Money and Finance, 152*, Article 103281. https://doi.org/10.1016/j.jimonfin.2025.103281

Karkowska, R., & Urjasz, S. (2025). *Volatility spillovers and hedging effectiveness of green bonds: A cross-regional analysis*. Advance online publication. https://doi.org/10.1007/s10614-025-11177-8

Kim, S., Olivares, R. O., Park, D., Tian, S., & Yang, S. (2026). How sovereign sustainable bond issuance shakes up the corporate sustainable bond market? Evidence from Asian markets. *Asian Economic Policy Review, 21*(1), 57–67. https://doi.org/10.1111/aepr.70008

Kukah, A. S. K. (2025). *Major global carbon emissions trading schemes*. Advance online publication. https://doi.org/10.1108/CI-07-2024-0208

Mahfooz, T., & Singh, P. (2026). *Market impact of India's sovereign green bond trading facilitation: A case study using event study methodology*. Advance online publication. https://doi.org/10.1177/22785337251413402

Malone, L. (2025). How Moody's acquisition made ESG mainstream in credit ratings. *Journal of Applied Accounting Research*. Advance online publication. https://doi.org/10.1108/JAL-12-2024-0381

Naeem, M. A., Ashraf, S., Karim, S., & Moussa, F. (2024). Green finance under stress: Unraveling the spillover effects of tail risk. *International Review of Economics & Finance, 93*, 225–236. https://doi.org/10.1016/j.iref.2024.03.026

Nevatia, V. (2026). Sovereign green bonds: Risk-mitigating sustainability instruments in emerging markets. *Energy Economics, 155*, Article 109173. https://doi.org/10.1016/j.eneco.2026.109173

Oude Lansink, A., & Kapelko, M. (2025). Do sustainability-linked bonds pay lower interest rates? *Corporate Social Responsibility and Environmental Management, 32*(4), 5337–5347. https://doi.org/10.1002/csr.3246

Owusu-Mante, S., Chi, K.H.-K., & Gopalakrishnan, T. (2025). *Sovereign green bonds in practice: Effectiveness as a climate finance instrument in emerging economies*. Advance online publication. https://doi.org/10.1080/14693062.2025.2586572

Pan, C., Sun, C., Zhang, Y., Li, Y., & Naeem, M. A. (2026). Climate change exposure and global sovereign credit risk. *Journal of International Financial Markets, Institutions and Money, 107*, Article 102238. https://doi.org/10.1016/j.intfin.2025.102238

Pedini, L. (2025). Are there other fish in the sea? Exploring the hedge, diversifier and safe-haven properties of ESG investments. *Studies in Economics and Finance*. Advance online publication. https://doi.org/10.1108/SEF-05-2023-0255

Poggensee, J. (2025). The pricing of sustainability-linked bonds on the primary and secondary bond markets. *Journal of Asset Management, 26*, 411–431. https://doi.org/10.1057/s41260-024-00390-z

Romagnoli, S., & Santini, A. (2025). Climate risk and sovereign debt: Country-level exposures and scarcity effects in green bonds. *Decisions in Economics and Finance*. Advance online publication. https://doi.org/10.1007/s10203-025-00554-5

Seghini, C. (2026). The macroeconomics of climate change and public debt sustainability: Mapping transition and physical risks through output growth, primary balances and real rates. *Journal of Economic Surveys*. Advance online publication. https://doi.org/10.1111/joes.70060

Sharma, S. E., & Babic, M. (2025). Introduction: The international political economy of green finance. *Competition & Change*. Advance online publication. https://doi.org/10.1177/10245294241310546

Syed, A. A. (2025). Does green stock immune to policy uncertainties? Examining connectedness between green stock, green bond and uncertainty indexes. *Benchmarking: An International Journal*. Advance online publication. https://doi.org/10.1108/BIJ-06-2024-0505

Tyson, L. D., & Weiss, D. (2025). Climate finance: Progress, challenges, and opportunities on the path to a sustainable planet. *California Management Review, 67*(2). https://doi.org/10.1177/00081256241296902

Vulturius, G., Maltais, A., & Forsbacka, K. (2024). Sustainability-linked bonds: Their potential to promote issuers' transition to net-zero emissions and future research directions. *Journal of Sustainable Finance & Investment, 14*(1), 116–127. https://doi.org/10.1080/20430795.2022.2040943

Sub-Sovereign Issuance: PPPs, Pipelines, and Credit Enhancement

Charbel Karam

1 Introduction

The climate transition is financed in advance of its being recorded in markets. Urban transport, district energy, water networks, storm drainage, social housing retrofits, waste systems, and local adaptation projects are planned and operated at the sub-sovereign level even when national governments dominate formal borrowing. This creates a persistent mismatch between where climate investment occurs and where capital market access is concentrated (World Bank & United Nations Capital Development Fund, 2024; OECD, 2023).

Recent global tracking reinforces the scale of the problem. The State of Cities Climate Finance 2024 concludes that urban climate finance remains far below what cities need and still contains major blind spots, especially for adaptation, waste, water, and wastewater. In other words, the financing gap is not simply a matter of more capital. It is also a matter of how projects are structured, aggregated, and made investable at the sub-sovereign level (Press-Williams et al., 2024).

That mismatch has consequences. Sovereign thematic bonds can provide a national funding envelope, but they do not address the local challenges of project preparation, tariff design, balance sheet weakness, or PPP risk allocation. Nor can credit enhancement by itself compensate for poor intergovernmental systems or weak disclosure. Sub-sovereign sustainable finance becomes reliable only when fiscal architecture, pipeline discipline, and market design reinforce one another. The chapter develops that argument in four steps: why issuance remains limited, what makes a pipeline investable, how credit enhancement should be used, and what a supportive national architecture looks like.

C. Karam (✉)
Le Cnam Liban, Byblos, Lebanon
e-mail: charbel.karam@isae.edu.lb

H. Jarrar et al. (eds.), *Sovereign Sustainable Finance*, Sustainable Finance,
https://doi.org/10.1007/978-3-032-33239-4_7

Another angle matters here. Local governments are where climate investment becomes concrete, yet they rarely control the full financing chain. National governments define borrowing rules, transfers, procurement norms, and in many cases the taxonomies used to label spending as green or resilient. That institutional layering explains why sub-sovereign sustainable issuance is not simply a capital-market problem. It is a coordination problem across fiscal frameworks, project preparation systems, and intergovernmental relations. OECD work on regions and cities and the World Bank and United Nations Capital Development Fund review of local climate finance instruments both underline the same point: market access improves when local financing is embedded in a credible national architecture rather than treated as a stand-alone municipal initiative (Chatry et al., 2025; World Bank & United Nations Capital Development Fund, 2024).

What slows issuance, then, is not the absence of climate need. It is the gap between investment responsibility and financing readiness. A city may be expected to modernize buses, upgrade drainage, retrofit public buildings, and expand water treatment, but still operate with volatile transfers, weak asset registers, and limited room to retain revenue. In that setting, even a well-structured sustainable bond framework cannot substitute for basic creditworthiness. Delmon (2024) is useful on this point because blended structures and guarantees work best when they support bankable projects rather than compensate for unresolved governance problems. The same logic underpins broader calls to mobilize debt markets for a just transition: capital will flow more consistently where project preparation and fiscal governance reduce execution risk before the financing search begins (Delmon, 2024; Robins et al., 2023).

2 Why Sub-Sovereign Sustainable Issuance Remains Limited

The main constraints are structural rather than thematic. Many local and regional governments lack predictable own-source revenues, operate under rigid borrowing controls, or depend heavily on transfers whose timing and composition are not well aligned with debt service. In that setting, even a worthy climate project struggles to become financeable because investors are not only buying the project story. They are buying the issuer's fiscal system, reporting quality, and legal authority to borrow or pledge revenues (World Bank & United Nations Capital Development Fund, 2024).

Scale is a second barrier. Sub-sovereign climate projects are often smaller, more heterogeneous, and less standardized than sovereign programs. A single city may bundle electric buses, drainage works, landfill remediation, and building retrofits, each with different cash-flow and procurement characteristics. That heterogeneity raises transaction costs and can undermine investor confidence unless issuers follow a repeated financing template. Capital markets reward routine. Sub-sovereign issuers often operate on a one-off basis.

OECD evidence suggests the market footprint remains modest and heavily conditioned by national market design. In its 2025 paper on mobilizing sustainable finance for regions and cities, the OECD notes that subnational governments accounted for only around a tenth of total sustainable bond issuance. That share is not trivial, but it is far below the investment responsibilities of local and regional governments. The gap signals that access conditions, not policy ambition alone, remain the binding constraint (Chatry et al., 2025).

Administrative capacity also matters. The same local governments that must identify and implement climate projects are often those with the thinnest legal, financial, and procurement teams. When staff turnover is high and financial reporting lags, the market's risk premium rises quickly. This is why the financing debate should begin with institutional readiness rather than with instrument branding. A city that cannot produce audited accounts, a multi-year capital plan, and transparent disclosure of contingent liabilities is not yet a strong bond issuer, regardless of how green its project list may be (Table 1).

Predictability is the first real credit enhancement. Investors and lenders want to know not only whether a local entity can borrow, but also whether the rules governing repayment are durable. That means audited financial statements, transparent debt

Table 1 Core prerequisites for sub-sovereign issuance

Prerequisite	Why it matters for issuance	Typical weakness in weaker markets
Predictable fiscal framework	Investors assess repayment capacity, transfer reliability, and borrowing rules before they assess labels	Opaque transfers, unstable own revenues, and uncertain borrowing authority
Bankable project pipeline	Repeatable preparation lowers transaction costs and improves credibility with lenders and investors	One-off projects with weak appraisal, fragmented documentation, and shifting priorities
Appropriate risk allocation	PPP structures and revenue models need risks assigned to the party best able to manage them	Political, demand, or construction risks are pushed into contracts without realistic mitigation
Targeted credit enhancement	Guarantees, intercepts, reserve funds, or pooled vehicles can solve specific tenor or credit constraints	Enhancement is used as a substitute for reform and creates hidden contingent liabilities
National enabling architecture	Standardised disclosure, guidance, and oversight help local issuers reach markets consistently	Local entities are left to negotiate legal, fiscal, and reporting rules transaction by transaction

Note. Source Author's own elaboration based on World Bank and United Nations Capital Development Fund (2024), OECD (2023), Chatry et al. (2025), and World Bank Group et al. (2017)

records, enforceable borrowing approvals, and a realistic view of contingent liabilities arising from public enterprises and concessions. The World Bank guidance on municipal PPPs and the broader debt-management guidance issued with the International Monetary Fund both make a related point: financing instruments are easier to scale when governments clarify who bears demand risk, land risk, construction risk, and termination risk before they approach the market (World Bank, 2019; World Bank & International Monetary Fund, 2014).

Labels do not soften that discipline. A sustainable bond, a climate facility, or a pooled green note is still assessed through familiar questions about legal security, reporting quality, and repayment sources. This is why some sub-sovereign issuers find it easier to begin with pooled or intermediary structures than with direct market access. Otek Ntsama et al. (2021) show that green issuance in low and middle-income settings remains shaped by capacity and market-depth constraints, while the World Bank survey on sovereign and public thematic bonds stresses the importance of transparent frameworks and post-issuance reporting. In practice, the sustainable label amplifies scrutiny because investors expect both financial soundness and credible environmental use of proceeds (Otek Ntsama et al., 2021; World Bank, 2022). The practical warning has become sharper in recent work on developing countries, which shows that green labels may add certification and reporting costs without solving the underlying credit and pipeline problem (Gorelick et al., 2024).

It is also useful to distinguish between general-purpose municipalities and ring-fenced utilities. Utilities with audited accounts, stable tariffs, and clearer cash flows may reach lenders more easily than municipalities whose finances are deeply intertwined with transfers and political cycles. That does not make utilities automatically stronger issuers, but it does suggest that sub-sovereign market development will rarely follow a single template. Countries may need different channels for cities, water utilities, transport agencies, and pooled financing vehicles. A realistic market architecture starts from those differences rather than assuming that all local entities can graduate to the same funding model at the same pace (World Bank & United Nations Capital Development Fund, 2024; OECD, 2023).

Intergovernmental design remains decisive even where project quality is strong. If transfer formulas are opaque, delayed, or politically unstable, markets will discount the issuer even when the climate project itself is compelling. That is why some countries advance local climate finance more effectively through pooled facilities or sovereign intermediated structures than through immediate stand-alone issuance. The practical issue is not prestige. The question is whether the repayment architecture is sufficiently dependable to support long-term borrowing without overpricing (Chatry et al., 2025; World Bank & International Monetary Fund, 2014).

3 Building a PPP and Investment Pipeline that Markets Can Trust

An investable pipeline starts before financing is discussed. It begins with a medium-term capital plan that ranks projects by strategic relevance, fiscal affordability, readiness, and expected climate impact. That sounds obvious, yet many local entities move directly from political commitment to financing search. Markets read that sequencing is a weak discipline. A credible issuer instead demonstrates that the project has already passed through basic screening on demand, land, environmental compliance, procurement path, and life-cycle cost.

PPP pipelines deserve even more discipline because financing is highly sensitive to risk allocation. A good PPP project is not defined by off-balance-sheet appearance or by the ability to defer public payment obligations. It is defined by whether construction, demand, operating, and political risks are assigned to the parties best able to manage them and whether long-term affordability has been stress-tested. The World Bank's Municipal PPP Framework is helpful here because it treats municipal PPPs as a governance challenge first and a financing challenge second. That ordering is exactly right (World Bank, 2019). The broader empirical literature on PPPs supports the emphasis on preparation and risk allocation. Reviews of infrastructure, health, and education PPPs show that outcomes depend less on the label itself than on contract design, information quality, governance, and enforcement capacity (Fabre & Straub, 2023).

Repeatability is a major advantage. Investors become more comfortable when they see a recognizable framework applied across successive transactions. Standardized feasibility criteria, consistent climate taxonomy use, and routine post-award disclosure lower information asymmetry. This is one reason why sub-sovereign bond markets are stronger in countries that have created programmatic issuance channels rather than relying solely on bespoke transactions. A pipeline is more than a list of projects. It is a production system for credible financing opportunities.

Mexico City's green bond experience illustrates the value of preparation. OECD reporting highlights that the city's green bond was oversubscribed by 2.5 times despite difficult market conditions. The point is not simply that investors liked the climate label. It is that the issuer had a credible financing framework, visible projects, and a broader story about urban investment needs. That is the standard sub-sovereign issuers should aim for: climate relevance embedded within conventional market credibility (OECD, 2023).

A credible PPP and investment pipeline starts well before procurement. The strongest local issuers usually work from a medium-term capital plan that ranks projects by readiness, affordability, developmental value, and climate contribution. That planning step sounds technical, yet it is where many sub-sovereign programs either become investable or fall apart. Markets respond more positively when feasibility work, land acquisition, environmental screening, demand analysis, and life-cycle costing are done in a disciplined sequence. OECD work on financing cities and Delmon's treatment of infrastructure finance both show that capacity for preparation is often the real bottleneck, not a lack of investor appetite (Delmon, 2024; OECD, 2023).

There is also a practical distinction that is often blurred in policy debate. Some local climate projects generate user-linked cash flows and may fit a PPP or utility-style financing model. Others are essential public investments whose value lies in avoided losses, public health, or inclusion rather than direct revenue. Flood defenses, informal-settlement drainage, and urban heat adaptation often belong in the second group. Trying to force those projects into PPP structures can create weak contracts and unrealistic revenue assumptions. Bhandary et al. (2021) remind us that climate finance instruments should be evaluated against effectiveness and fairness, not only mobilization volume. Mazzucato (2016) similarly argues that public financing structures need to be aligned with the public mission rather than imitating private project logic. That distinction is especially important for sub-sovereign climate pipelines, where fiscal realism matters more than branding (Bhandary et al., 2021; Mazzucato, 2016).

Project preparation facilities can make a significant difference because they solve a recurring problem rather than a single-deal problem. Local entities often know what they want to build, yet lack the resources to commission feasibility work, standardize contracts, or structure procurement in ways financiers trust. A modest preparation platform that supplies technical, legal, and financial support can therefore do more for market access than a large but sporadic subsidy. It reduces transaction costs across the pipeline and helps create comparable projects that can be pooled or funded through repeat programs (Delmon, 2024; World Bank, 2019).

Standardization matters for climate projects, especially because many are medium sized rather than large. Electric bus fleets, district cooling upgrades, building retrofits, waste improvements, and local flood works are often too small on their own to attract efficient market attention. Pooled procurement, common documentation, and shared reporting can make these investments legible to investors without forcing them into unrealistic mega-project structures. That kind of aggregation is one of the few practical routes from policy ambition to investable local climate portfolios (OECD, 2023; World Bank & United Nations Capital Development Fund, 2024).

4 Credit Enhancement and Pooled Structures

Credit enhancement is often the bridge between local investment need and market access. Guarantees, reserve funds, intercept mechanisms, subordinated tranches, liquidity facilities, and pooled borrowing arrangements can reduce credit risk enough to attract investors who would not otherwise purchase municipal or utility paper. Used well, these tools compensate for market immaturity, not for fiscal denial. The difference matters. Enhancement should support sound issuers and viable projects, not shield structurally weak balance sheets from scrutiny (World Bank & United Nations Capital Development Fund, 2024; Delmon, 2024).

Development bank guarantees are especially useful where a long tenor is essential. Climate-smart water, transport, and resilience projects often require maturities that domestic lenders are reluctant to provide. A partial guarantee can narrow that gap and allow a transaction to reach closure while preserving incentives for prudent issuer behavior. World Bank examples in local government climate finance show how guarantees have been used alongside PPP structures and municipal borrowing to mobilize infrastructure finance in developing-country settings (World Bank & United Nations Capital Development Fund, 2024).

Pooled structures deserve more attention than they usually receive. Many smaller municipalities will never be efficient direct issuers because their borrowing needs are episodic and too small to sustain regular investor relations or benchmark liquidity. Pooled financing facilities or intermediary issuers can aggregate projects, standardize disclosure, and spread fixed transaction costs. They can also support shared technical assistance on climate appraisal and reporting. The trade-off is governance: once projects are pooled, selection and monitoring rules must be especially clear.

Enhancement should also be sequenced carefully. A country that uses guarantees before it has clarified subnational borrowing rules or transfer predictability may crowd in capital temporarily while deepening moral hazard. Better results come when enhancement is paired with stronger disclosure, multi-year capital planning, and explicit rules on how contingent liabilities are approved and reported. In other words, credit enhancement works best when it amplifies reform rather than replaces it.

Credit enhancement works when it changes a specific constraint rather than vaguely signaling support. Partial credit guarantees can improve tenor or lower spreads. Intercept mechanisms can strengthen repayment confidence where transfer systems are reliable. Reserve funds and subordinated tranches can make pooled facilities more investable for institutional investors. The World Bank and United Nations Capital Development Fund review provides a useful cross-country lesson here: local climate finance deepens when enhancements are tied to transparent eligibility, reporting, and monitoring rules rather than negotiated, case-by-case arrangements. In other words, enhancement is most effective when it standardizes access rather than creating one-off exceptions (World Bank & United Nations Capital Development Fund, 2024).

That said, guarantees should never become a device for hiding risk outside the budget. Once national governments begin backstopping municipal or project borrowing, they are taking on contingent liabilities that belong inside fiscal oversight. The older debt-management guidance remains relevant for exactly this reason. Governments need to disclose guarantees, cap exposures, assess expected fiscal cost, and decide when a credit enhancement supports market development and when it simply socializes poorly allocated risk. The difference is crucial. Sustainable finance gains credibility when guarantees are explicit, priced where appropriate, and paired with stronger local governance rather than used as a substitute for it (World Bank & International Monetary Fund, 2014, 2019).

Pooled finance deserves special attention because it can address problems of scale, credit, and reporting at once. Where many local issuers are individually too small or too weak for direct market access, a pooled facility can diversify exposure, standardize documents, and create a more visible issuance program. The climate-finance advantage is that it can also align project selection and reporting across a wider set of local investments. Yet the facility must be designed carefully. Weak entry criteria or politically motivated participation can simply turn pooling into collective weakness. Strong gating rules are what make diversification valuable rather than cosmetic (World Bank & United Nations Capital Development Fund, 2024; Chatry et al., 2025).

Local-currency financing is another issue that deserves more prominence. Climate and urban infrastructure assets are long-lived and often supported by local-currency revenues or budget flows. Financing them through mismatched currency structures can create vulnerabilities that overwhelm any environmental benefit from the investment itself. This is one reason why national development banks, guarantee facilities, and domestic capital-market development remain relevant to the sub-sovereign agenda. The financing question is not only whether money can be raised, but whether it can be raised in a structure that the local balance sheet can survive (Delmon, 2024; Robins et al., 2023).

5 Designing a National Market Architecture for Local Climate Finance

Sustainable sub-sovereign finance scales only when national policy creates a usable operating environment. That environment includes borrowing legislation, fiscal rules, transfer design, disclosure standards, and institutional support for project preparation. Countries that leave municipalities and utilities to invent these elements on a transaction-by-transaction basis usually end up with thin issuance, repeated advisory costs, and limited investor learning. A market architecture approach is more durable because it reduces uncertainty for both issuers and creditors (Chatry et al., 2025; OECD, 2023).

The national role is not to centralize every transaction. It is to standardize the conditions under which decentralized financing works. That may include national PPP guidance, climate-related disclosure templates, pooled issuance windows, guarantee frameworks, or fiscal incentives for investment-grade borrowing practices. The OECD's work on financing cities and on subnational sustainable finance points to exactly this kind of enabling architecture, with examples from India's federal incentives for municipal bond issuance and the growth of local SDG-linked borrowing in Japan (Chatry et al., 2025; OECD, 2023).

A second pillar is fiscal transparency across levels of government. Green budgeting and climate-related expenditure tracking are often discussed only at the central government level, yet they are highly relevant to local capital planning because they help connect projects, grants, debt, and outcomes. When local pipelines are aligned to multi-year climate expenditure frameworks, financing choices become easier to justify and monitor. When they are not, labeled sub-sovereign borrowing can drift into disconnected project marketing.

The strategic conclusion is straightforward. Sub-sovereign sustainable issuance should be treated as a system-building exercise. Where institutional fundamentals are weak, sovereign on-lending, pooled structures, or guarantee-backed financing may be the right early stage. Where local fiscal systems are stronger, direct market issuance can scale. In both cases, the objective is the same: build a pipeline and a governance architecture that allow climate finance to move reliably from national ambition to local delivery.

The missing middle in sub-sovereign sustainable finance is institutional architecture. Countries that want deeper local issuance need more than a few pilot transactions. They need common taxonomies, reporting templates, project-preparation support, and predictable rules that link local borrowing to national fiscal oversight. This is where sovereign and sub-sovereign agendas meet. OECD guidance on green budgeting and the World Bank review of sovereign thematic bonds both suggest that common definitions and reporting disciplines reduce fragmentation and make it easier to connect budgets, capital plans, and market disclosure. When that happens, local issuers no longer have to reinvent sustainable finance one transaction at a time (OECD, 2021, 2024; World Bank, 2022).

Country experience points in the same direction. Owusu-Mante et al. (2025) show that sovereign green programs in Chile, Egypt, Indonesia, and Nigeria have value not only because they raise money, but because they push governments to define eligible spending, reporting practices, and investor communication routines more clearly. Those routines can serve as enabling infrastructure for sub-sovereign actors, especially when implementation is decentralized. Still, the goal is not to force every city or utility to issue stand-alone bonds. In many cases, the better outcome is a layered system in which national borrowing, local investment planning, pooled facilities, and targeted guarantees work together. What matters is dependable local delivery under rules that are fiscally honest, transparent, and scalable (Owusu-Mante et al., 2025; Chatry et al., 2025; Robins et al., 2023).

6 Conclusion

Sub-sovereign sustainable finance is difficult because it sits at the intersection of local delivery and capital-market discipline. Climate investment needs are highly visible at the city and regional level, but market access depends on fiscal predictability, legal clarity, disclosure, and project preparation. Labels alone do not overcome those requirements.

The most robust path is sequential. First, build the fiscal and project architecture. Then use credit enhancement and pooled structures where necessary. Finally, deepen direct market access for issuers that can sustain it. PPPs, municipal bonds, and sustainability-linked local borrowing all have a role, but only when they rest on disciplined pipelines and transparent intergovernmental systems. That is how sub-sovereign finance becomes scalable rather than episodic.

Two sequencing lessons stand out from the literature. First, local issuers need clean fundamentals before they need thematic branding. That means reliable accounts, transparent debt data, a realistic capital plan, and workable procurement discipline. Second, market development is usually staged. Pooled facilities, guarantee-backed borrowing, or sovereign on-lending can build performance records before standalone local issuance is attempted. This staged approach is consistent with the World Bank and United Nations Capital Development Fund review and with OECD work on cities and regions, both of which emphasize institutional progression rather than one-step market access (World Bank & United Nations Capital Development Fund, 2024; Chatry et al., 2025).

In practice, the most successful national strategies will probably be those that stop treating sub-sovereign sustainable finance as a niche product line. It is better understood as the financing layer of multilevel climate governance. Once that point is accepted, the agenda becomes clearer: prepare projects early, assign risks honestly, use credit enhancement sparingly and transparently, and connect local investment plans to national budget and debt frameworks. That is less glamorous than announcing a new label, but it is far more likely to produce durable market access and bankable climate delivery at the local level (Delmon, 2024; OECD, 2023; World Bank, 2022).

Sovereign and local strategies, therefore, have to be joined up more deliberately than they often are. National thematic bond programs, green budgeting reforms, or sovereign climate frameworks can create useful standards, but they will not automatically solve local execution constraints. Those standards need to be connected to local investment planning, guarantee policy, pipeline support, and reporting methods. Once that connection is made, the boundary between sovereign and sub-sovereign sustainable finance becomes less a hard division and more a financing chain.

The most credible end state is not necessarily a large number of municipal green bonds. In some countries, it will be a mixed ecosystem of sovereign borrowing, pooled local facilities, utility finance, development-bank intermediation, and carefully chosen PPPs. That mixed architecture is often a sign of maturity, not incompleteness, because it recognizes that local climate delivery needs several financing channels rather than a single market template.

That sequencing mindset is what makes the sub-sovereign chapter state-of-the-art rather than aspirational. It does not assume that local climate finance will mature everywhere through the same headline instrument. It assumes instead that credible market access grows out of stronger institutions, better projects, and more honest risk allocation.

Put simply, sub-sovereign sustainable finance works best when market design follows public investment reality rather than trying to force public investment into a single market design.

References

Chatry, I., Wheeler, C., Kimura, Y., & Lafitte, C. (2025). *Mobilising sustainable finance for regions and cities* (OECD Regional Development Papers No. 132). OECD Publishing. https://doi.org/10.1787/2af8dadb-en

Bhandary, R. R., Gallagher, K. S., & Zhang, F. (2021). Climate finance policy in practice: A review of the evidence. *Climate Policy, 21*(4), 529–545. https://doi.org/10.1080/14693062.2020.1871313

Delmon, J. (2024). Blended finance. In *Innovative funding and financing for infrastructure: Addressing scarcity of public resources* (pp. 176–196). Cambridge University Press. https://doi.org/10.1017/9781009340182.011

Fabre, A., & Straub, S. (2023). The impact of public–private partnerships (PPPs) in infrastructure, health, and education. *Journal of Economic Literature, 61*(2), 655–715. https://doi.org/10.1257/jel.20211607

Gorelick, J., Cara, E., & Kavoo, G. (2024). The fallacy of green municipal bonds in developing countries. *World, 5*(4), 929–951. https://doi.org/10.3390/world5040047

Mazzucato, M. (2016). From market fixing to market-creating: A new framework for innovation policy. *Industry and Innovation, 23*(2), 140–156. https://doi.org/10.1080/13662716.2016.1146124

OECD. (2021). *Green budget tagging: Introductory guidance and principles.* OECD Publishing. https://doi.org/10.1787/fe7bfcc4-en

OECD. (2023). *Financing cities of tomorrow: G20/OECD report for the G20 Infrastructure Working Group under the Indian Presidency.* OECD Publishing. https://doi.org/10.1787/51bd124a-en

OECD. (2024). *Green budgeting in OECD countries 2024.* OECD Publishing. https://doi.org/10.1787/9aea61f0-en

Otek Ntsama, U. Y., Yan, C., Nasiri, A. et al. (2021). Green bonds issuance: Insights in low-and middle-income countries. *International Journal of Corporate Social Responsibility, 6*, 2. https://doi.org/10.1186/s40991-020-00056-0

Owusu-Mante, S., Chi, K.H.-K., & Gopalakrishnan, T. (2025). Sovereign green bonds in practice: Effectiveness as a climate finance instrument in emerging economies. *Climate Policy,* 1–15. https://doi.org/10.1080/14693062.2025.2586572

Press-Williams, J., Negreiros, P., Fernandes, P. de A., Meattle, C., Abdullah, H., Vieira, A., Diaz, J., & Melling, B. (2024). *The state of cities climate finance 2024.* Cities Climate Finance Leadership Alliance and Climate Policy Initiative.

Robins, N., Curran, B., Plyska, O., Burge, L., & Van Coppenolle, M. (2023). Mobilising global debt markets for a just transition. Grantham Research Institute on Climate Change and the Environment, London School of Economics and Political Science.

World Bank. (2019). *Municipal public-private partnership framework: Guidance note.* World Bank.

World Bank. (2022). *Sovereign green, social and sustainability bonds: Unlocking the potential for emerging markets and developing economies.* World Bank.

World Bank, & International Monetary Fund. (2014). *Revised guidelines for public debt management.* World Bank.

World Bank, & International Monetary Fund. (2019). *Developing a medium-term debt management strategy (MTDS): Updated guidance note for country authorities.* World Bank and International Monetary Fund.

World Bank, & United Nations Capital Development Fund. (2024). *Local governments climate finance instruments: Global experiences and prospects in developing countries.* World Bank.

World Bank Group. (2016). *World Bank Group guarantee products: Guidance note.* World Bank Group.

World Bank Group, Asian Development Bank, European Bank for Reconstruction and Development, Global Infrastructure Hub, Inter-American Development Bank, Islamic Development Bank, OECD, United Nations Economic Commission for Europe, & United Nations Economic and Social Commission for Asia and the Pacific. (2017). *Public-private partnerships reference guide (Version 3).* World Bank Group.

HR and Governance in Debt Offices: Capabilities and Operating Models

Enis Mulolli and Xhavit Islami

1 Introduction

As financial markets become increasingly volatile and interconnected, the operational environment of debt offices has grown more complex, requiring sophisticated governance mechanisms and highly specialized human resource capabilities to manage risks, coordinate stakeholders, and sustain long-term financial stability. Debt offices play a pivotal role in maintaining fiscal sustainability and ensuring global financial stability. The effective debt management extends beyond macroeconomic strategy to encompass the legal and managerial governance structures that shape organizational operations (Fukuyama, 2016; World Bank, 2021), as well as the strategic alignment of human capital assets with institutional objectives (Kaehler & Grundei, 2018). The establishment of international standards such as ISO 30408 in 2016 has further underscored the importance of "human governance" in financial institutions, framing HR governance not merely as an administrative function but as a core mechanism for optimizing performance, mitigating enterprise risk, and enabling executive decision-making (ISO, 2016; Mercer Inc., 2003).

Despite the growing body of research on governance, debt management, and human resource management, the intersection of HR and governance within debt offices remains fragmented and insufficiently explored. Much of the existing literature has predominantly concentrated on macroeconomic dimensions of debt management, such as governance and public debt (Tarek & Ahmed, 2017a), institutional quality (Mehmood et al., 2022), economic growth (Yusuf & Mohd, 2024), while the organizational and human-capability dimensions of debt management have received comparatively limited attention. Therefore, despite these contributions, a deeper

E. Mulolli · X. Islami (✉)
Faculty of Economics, AAB College, Pristina, Kosovo
e-mail: xhavit.islami@aab-edu.net

E. Mulolli
e-mail: enis.mulolli@aab-edu.net

H. Jarrar et al. (eds.), *Sovereign Sustainable Finance*, Sustainable Finance,
https://doi.org/10.1007/978-3-032-33239-4_8

understanding of the interrelationships between HR and Governance in Debt Offices is essential, particularly in relation to collaboration networks and the evolution of the field, such as co-citation, and co-occurrence networks along timelines. To address these gaps, this study systematically analyzes the existing literature on HR and governance research within debt offices, to uncover its intellectual structure, key interconnections, evolutionary trajectories, predominant research themes, and to propose future investigation on the topic. Thus, two guiding research questions are addressed: (1) What are the predominant research themes, scholarly contributions, and developmental trajectories that characterize HR and governance in debt offices? (2) Where is the research on HR and governance capabilities in debt offices heading, and what promising research avenues for further investigation? To investigate these questions, we employed a combined bibliometric-systematic literature review of 132 journal articles published between 2004 and 2025. Using performance analysis, co-citation analysis, bibliographic coupling, and co-occurrence analyses, the review maps the intellectual structure, identifies the main thematic clusters, and the dominant developments and emerging research areas in the field.

The chapter is organized as follows. Section 2 presents definitions of the key concepts and their interrelationships. Section 3 describes the research methodology used to collect the data and the analytical procedures adopted. Section 4 presents the findings of the literature review, discusses the main challenges, and highlights opportunities for future research, while Sect. 5 concludes the study and outlines its implications for theory and practice.

2 Conceptual Background

Public debt management is described as "the process of establishing and executing a strategy for managing the government's debt in order to raise the required amount of funding at the lowest possible cost over the medium to long run, consistent with a prudent degree of risk" (Lewis & Viñals, 2014, p. 11). On other hand, a "Debt Office can be described as an agency with at least some autonomy from the political sphere, which is responsible for the management of the State debt" (Magnusson, 1999, p. 2). The role of a debt office "depends on its debt management functions, such as the duties that laws and regulations assign to it, on the strategy it chooses to adopt and, on the means, placed at its disposal to achieve its goals" (Borresen & Cosio-Pascal, 2002, p. 1).

2.1 Governance in Debt Offices

The term "governance" does not have a universally settled definition and carries at least three primary meanings (Fukuyama, 2016). First, it refers to international cooperation via non-sovereign bodies operating outside state systems. Second, governance serves as a synonym for public administration, emphasizing the effective implementation of state policies. Third, it denotes the regulation of social behavior through networks and other non-hierarchical mechanisms (Fukuyama, 2016). "Governance is about establishing, promoting and supporting a specific type of relationship between governmental and non-governmental actors in the governing process" (Howlett & Ramesh, 2014, p. 318). In the context of debt management, the term "governance" refers to the legal and managerial structure that shapes and directs the operations of government debt management, including accountability, strategy implementation, and reporting arrangements (World Bank, 2021). Accordingly, governance frameworks can be understood as the outcome of interactions among multiple actors, including external stakeholders such a legislator, a public code commission or an ISO committee, to create norms that are applicable in organizations (Kaehler & Grundei, 2018).

2.2 Human Resource in Debt Offices

Armstrong (2008) defines human resource management (HRM) as a strategic, coherent access of management the most valuable asset of an organization-people that work there individually or collectively contribute to achieve the objectives of the organization. To comprehensively understand the impact of HRM on organizational outcomes, HRM practices should be analyses from different perspectives, highlighting their differences (Mulolli et al., 2025), as well as their synergistic effects when these practices complement one another (Islami & Mulolli, 2021). On the other hand, the linkage between human resources and governance is relatively recent as a conceptual development. In one of the pioneering contributions to this field Mercer Inc. (2003, p. 2) defines HR governance as the act of leading the HR function and managing related investments to: (i) optimize performance of the organization's human capital assets; (ii) fulfill fiduciary and financial responsibilities; (iii) mitigate enterprise HR risk; (iv) align the function's priorities with those of the business; and (v) enable HR executive decision making. The first international ISO standards for human resource management were published in 2016, notably including ISO 30408, which is dedicated exclusively to governance aspects, this standard addresses HR governance, it refers to it as "human governance", allowing the terms to be used synonymously (Kaehler & Grundei, 2018). By shifting the focus of HR governance towards the HRM system, the understanding of HR governance is broadened, truly integrating all people management issues (Kaehler &Grundei, 2018). In the context of debt offices, the application of HR governance principles strengthens management oversight by ensuring that HR systems are aligned with organizational risk management frameworks, strategic objectives, and the development of appropriate capabilities and operating models.

3 Methodology

3.1 Data Collection and Search Strategy

An initial scan of the literature was conducted to develop an overview of the research field and to identify a set of representative keywords for the database search. To ensure methodological rigor, Scopus was selected as the primary data source due to its extensive coverage of peer-reviewed journals in business and management (Ferasso et al., 2025), economics, econometrics and finance, and its recognized reliability for bibliometric and network analyses (Baas et al., 2020). A search query was defined the conducted search was in the TITLE-ABS-KEY field, on Scopus. To refine the results, the search was limited to journal articles and review papers published in English within the subject areas of "Economics, Econometrics and Finance" and "Business, Management and Accounting, this search yielded a total of 249 documents indexed in Scopus database. To ensure the relevance of the collected literature, the sampled articles were independently screened by two researchers. Finally, 132 articles were included in the review, and their metadata were retrieved from Scopus. The PRISMA flow diagram illustrating the steps of literature identification and search refinement is presented in Fig. 1. We conducted descriptive statistical, bibliometric, and network analyses using the bibliometrix package in R (Aria &Cuccurullo, 2017), VOSviewer (van Eck & Waltman, 2016).

4 Findings and Discussion

4.1 Performance Analysis

The final dataset consisted of 132 publications produced by 272 authors and disseminated across 97 distinct sources. In the HR/Governance in Debt field, annual publication growth reached 19.2%, with an average paper age of 5.28 years; notably, 24.24% of studies featured international co-authorship (see Table 1). Figure 2 depicts the trajectory of scientific publications, serving as a key indicator of research activity and emerging trends in the field. The dataset's earliest publication dates to 2004, spanning continuously through 2025. From that point onward, the analysis covers publications up to the year 2025. While publication activity remained relatively modest in the initial years, a substantial upward trend becomes evident in the later period. In particular, a pronounced surge occurs from 2024 onward, with 15 articles published in that year, followed by a significant peak in 2025, reaching 40 publications, underscoring the field's accelerated development.

As noted above, the 132 analyzed articles appeared across 97 distinct journals. The top five contributing journals were Finance Research Letters (9 articles), Applied Economics (6), Empirical Economics (4), Review of International Political Economy

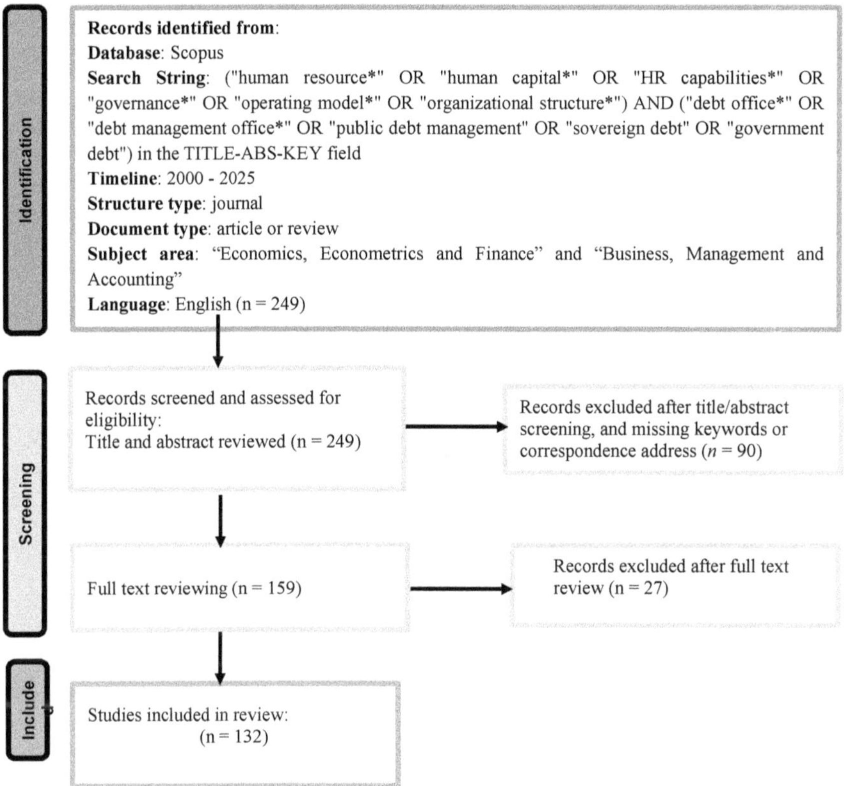

Fig. 1 PRISMA flow chart. *Source* PRISMA 2020 statement (Page et al., 2021)

(3), and Journal of Common Market Studies (2). In terms of authorship, 93.75% of the 272 scholars (255 authors) contributed just one article, while 17 authors (6.25%) published two or more. Among these, B.A, Tarek, S. Li, Y. Wang, and Z. Wang emerged as the most prolific, each with three articles (Table 2).

The most cited article, by Featherstone (2011), garnered 275 global citations, affirming its substantial influence and enduring relevance. Ranked second and third were Tarek and Ahmed (2017a, 2017b) and Ball (2012), each with 47 citations. Overall impact patterns show that 34 articles (25.8%) have achieved moderate to high visibility with more than 10 citations each. Conversely, 30 papers (22.7%) remain uncited, a pattern largely attributable to recency bias, as the majority of these were published in 2025 and have had insufficient time to accumulate scholarly attention (Table 3).

The word cloud in Fig. 3 emphasizes keywords such as "governance approach", "debt", "government debt", "fiscal policy" and "debt crisis" reflect the main focus of the field of HR and governance within debt offices.

Figure 4 presents a three-field Sankey diagram mapping the interconnections among countries, keywords, and journals, where box dimensions reflect frequency

Table 1 Main information about the data

Category	Detail	Results
Data main information	Timespan	2004:2025
	Sources (Journals)	97
	Documents	132
	Annual Growth Rate %	19.2
	Document Average Age	5.28
	Average citations per doc	9.583
	References	1157
Document content	Keywords Plus (ID)	223
	Author's Keywords (DE)	515
Authors	Authors	272
	Authors of single-authored docs	33
Authors collaboration	Single-authored docs	33
	Co-Authors per Doc	2.22
	International co-authorships %	24.24
Document types	Article	126
	Review	6

Source Authors' elaboration using Biblioshiny package in R

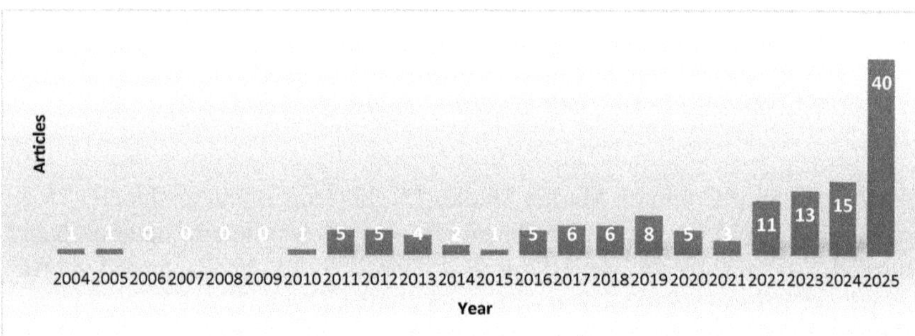

Fig. 2 Annual scientific production in the field

of occurrence. China dominates the landscape with the highest publication volume, with its scholars concentrating predominantly on "government debt" and "local government debt." The keyword "public debt" shows a distinct journal affiliation, appearing primarily in Economies. Meanwhile, "government debt" also garners substantial attention from U.S. and German researchers, indicating transnational thematic convergence despite geographic clustering. This analysis proves instrumental in pinpointing dominant trends and pivotal contributions, while illuminating probable trajectories for future research (Kraus et al., 2024). Furthermore, identifying the leading journals and which authors are most prolific and frequently cited helps

Table 2 Ranking of the five most influential journals in the field

Journal	Number of articles	Total citations (TC)	Publication year start
Finance Research Letters	9	60	2023
Applied Economics	6	24	2014
Empirical Economics	4	29	2017
Review of International Political Economy	3	19	2025
Journal of Common Market Studies	2	299	2011

Source Authors' elaboration using Biblioshiny package in R

Table 3 Ranking of the five most cited articles in the field

Article	Journal	Total citations (TC)	Total citations per year
Featherstone (2011)	Journal of Common Market Studies	275	17.18
Tarek and Ahmed (2017a, 2017b)	Economic Analysis and Policy	47	4.7
Ball (2012)	Public Money and Management	47	3.13
Coles et al. (2012)	Journal of Destination Marketing and Management	29	1.93
Li and Qi (2023)	Finance Research Letters	28	7

Source Authors' elaboration using Biblioshiny package in R

Fig. 3 WordCloud. *Source* Authors' elaboration using Biblioshiny package in R

inform theoretical foundations and define how the field is consolidating its intellectual core, key outlets and influences in HR governance for debt offices, guiding collaborations and future agendas. Furthermore, identifying the leading journals and most prolific, highly cited authors clarifies the field's theoretical foundations and intellectual core in HR and governance for debt offices. These insights underscore gaps warranting targeted research and help define future research agendas.

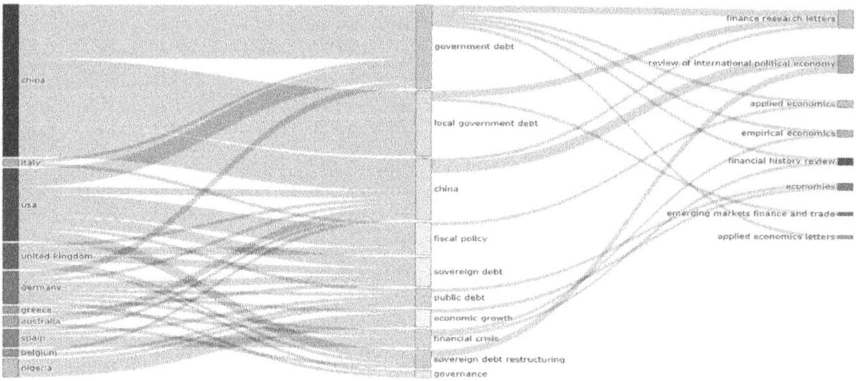

Fig. 4 Three-field plot (Sankey diagram) depicting linkages among countries (left), keywords (center), and journals (right). *Source* Authors' elaboration using Biblioshiny package in R

4.2 Co-citation Analysis

Co-citation analysis assesses similarity among articles, authors, or journals based on how often two items are cited together by one or more subsequent works (Dabic et al., 2015). Co-citation analysis evaluates the references cited by the publications included within the dataset (Caputo et al, 2021) and offers a dynamic representation of a topic from the past, rather than the present or the future (Caputo et al., 2019). In this regard, co-citation analysis serves as a powerful methodological tool to investigate the intellectual foundations of research on HR and governance in debt contexts through the analysis of references cited by papers in our dataset. By performing this analysis, we were able to identify the 5 most linked papers, and cited authors and journals that constitute the theoretical pillars of the research. The results on Table 4 show how such research is grounded in the literature about governance and debts (Alesina & Tabellini, 1990; Arellano, 2008; Battaglini & Coate, 2008).

4.3 Bibliographic Coupling

Bibliographic coupling occurs when two articles cite a common third source, indicating that the two articles potentially discuss a common topic (Kessler, 1963). Rather than forming clusters, we employed this method to assess network relevance among papers, authors, and journals (Caputo et al., 2023). Our analysis encompassed all 132 papers, 272 authors, and 97 journals from the dataset (Table 5). Coupling strength was quantified using VOSViewer's total link strength algorithm, which measures similarity and interconnectedness independent of citation counts (van Eck & Waltman, 2016). This approach revealed the five most connected papers, authors, and journals shaping research on HR and governance in debt contexts. The findings indicate a

Table 4 Co-citation analysis

Cited references	Link strength	Cited author	Citations	Citated journal	Citations
Alesina and Tabellini (1990)	49	Alesina, A	17	Review of financial studies	8
Arellano (2008)	40	Acharya, V	13	Quarterly journal of economics	8
Battaglini and Coate (2008)	36	Alt, J. E	5	American economic review	8
Chen et al. (2020)	36	Barro, R. J	4	Review of economic studies	7
Acemoglu et al. (2001)	35	Bovens, M. A	4	Journal of financial economics	5

Source Authors' elaboration using Vosviewer

Table 5 Bibliographic coupling analysis

Articles	link strength	Cited author	Citations	Citated journals	Citations
Tarek and Ahmed (2017a)	29	Featherstone, K	275	Journal of Common Market Studies	299
Tarek and Ahmed (2017b)	29	Tarek, B. A	87	Finance Research Letters	60
Ben Ali et al. (2019)	26	Ahmed, Z	67	Journal of financial stability	51
Mehmood et al. (2022)	19	Ball, I	47	Public money and management	49
Yusuf and Mohd (2024)	18	Coles, T. E	29	Economic analysis and policy	47

Source Authors' elaboration using Vosviewer

field converging toward coherent core contributions while maintaining openness to new developments and perspectives.

4.4 *Co-occurrence of Keywords*

The keyword co-occurrence analysis reveals the field's intellectual structure by identifying and grouping its core research topics, serving as a valuable complement to the previous analyses. The *.CSV file was analyzed using VOSviewer (version 1.6.20), following the methodological framework outlined by Van Eck and Waltman (2023). A minimum threshold of two occurrences per keyword was set as the threshold (Jobstreibizer et al., 2025). From the 515 identified keywords, 62 keywords met the

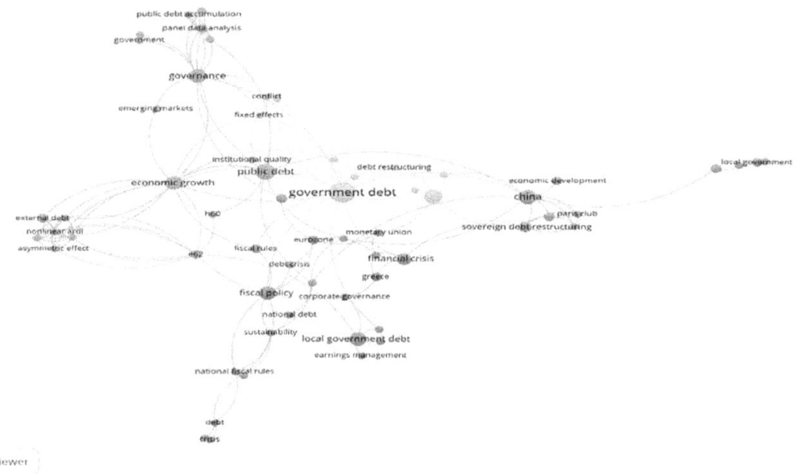

Fig. 5 Co-occurrence of keywords. *Source* Authors' elaboration using Vosviewer

threshold. As a result, the visualized co-occurrence network comprised 62 connected keywords and 141 links between them, forming four clusters (Fig. 5).

The keyword co-occurrence analysis revealed four distinct thematic clusters, each visualized through unique colors: green, red, yellow, and blue. The red cluster prominently features keywords such as local government debt, fiscal policy, financial crisis, and corporate governance. This concentration underscores the critical role of robust fiscal governance mechanisms and institutional oversight in effectively managing public debt. From the perspective of debt management offices, this stream emphasizes the need for strong governance frameworks and skilled personnel capable of implementing fiscal rules, managing risks, and ensuring transparency in debt operations. The green cluster includes keywords such as economic growth, governance and public debt. This stream of research examines how debt levels influence macroeconomic outcomes, from GDP trajectories to fiscal stability thresholds. For debt offices, this underscores the need for specialized HR talent skilled in econometric modeling and scenario analysis, enabling robust evaluation of borrowing strategies' sustainability and alignment with broader economic governance frameworks. The blue cluster centers on sovereign debt restructuring, economic development, China, and global governance, with a particular emphasis on international debt negotiations and restructuring mechanisms. These complex processes demand specialized expertise within debt management offices, including advanced negotiation capabilities, legal proficiency, and effective coordination with international financial institutions. The yellow cluster contains keywords such as debt restructuring, government debt, sovereign debt and employment. The literature here examines how governments navigate debt obligations and the broader economic repercussions, including labor market effects. For debt management offices, this underscores the need for strong institutional capacity, policy coordination, and skilled professionals who can manage

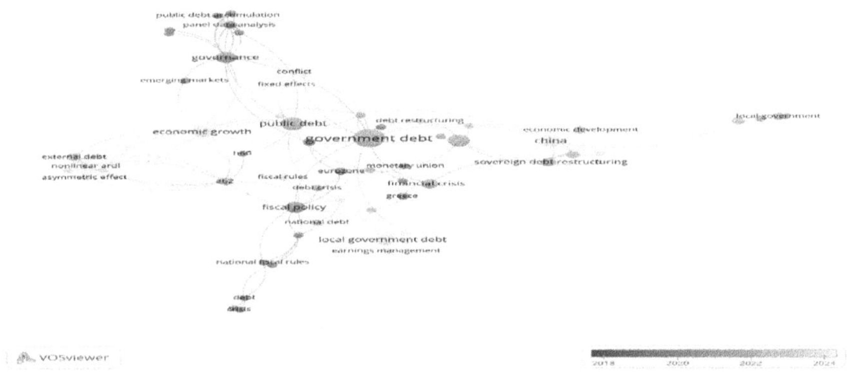

Fig.6 Co-occurrence network of keywords evolution. *Source* Authors' elaboration using Vosviewer

restructuring while balancing macroeconomic stability and social outcomes. While the timeline distribution of keywords (Fig. 6), highlights that future studies can contribute to a deeper understanding on topics economic growth, local government debt, fiscal rules, fixed effects, and international monetary fund, particularly in relation to the study's focus on HR and governance capabilities within debt management offices.

5 Conclusion

When it comes to HR and Governance within Debt Offices, most of the work lies ahead. Yet this is precisely what makes their intersection so valuable. Scholarly interest in this emerging domain has surged in recent years and will continue to grow. This research aimed to provide a thematic exploration and mapping of the developments at the intersection of HR and Governance within Debt Offices, offering a theoretical foundation, and outlining a future research agenda. By bridging macro-level indicators with micro-level dynamics, it provides a more integrated perspective for understanding and advancing HR and governance practices in debt offices. This study conducted a bibliometric and systematic review, by analyzing the Scopus dataset of 132 articles, with co-citation analysis, bibliographic coupling analysis, co-occurrence analysis, and timeline view analysis. Findings revealed a growing interest from scholars since 2022 and the peak of publications in 2025 alone accounting for 40 publications. Co-citation analysis identifies the theoretical pillars of the field, while bibliographic coupling reveals a convergence toward a coherent core of contributions, while maintaining openness to new developments and perspectives. Furthermore, by exploring the keyword co-occurrence network, four thematic clusters were formed, shown the importance of the interlinkages between HR and Governance within Debt Offices. Researchers are interested in (1) local government debt, fiscal policy, financial crisis, and corporate governance; (2) economic growth, governance and public debt; (3) sovereign debt restructuring, economic development, China, and global governance; (4) debt restructuring, government debt, sovereign debt and employment. While the timeline distribution of keywords suggests that future research should deepen insights into economic growth, local government debt, fiscal rules, fixed effects, and international monetary fund, particularly in relation to the study's focus on HR and governance capabilities within debt management offices. The insights emerging from this review indicate that HR is no longer a peripheral administrative function but a core governance mechanism. This is consistent with the notion that HR governance itself is a relatively new and still-evolving concept (Kaehler & Grundei, 2018), gaining clearer structure only in recent years through standards such as ISO 30408 (ISO, 2016).

Theoretical Implications—This study makes several key theoretical contributions to HR and governance literature in debt offices. It synthesizes up-to-date insights from economics, public administration, and management studies into a coherent theoretical landscape and evolutionary perspective on the field. The analysis demonstrates that while debt management has traditionally been theorized through macroeconomic or legal lenses, the HR and governance dimensions constitute a distinct and emergent domain requiring interdisciplinary theoretical development. Second, the co-occurrence network analysis identifies both dominant and emerging research themes, illustrating a temporal evolution. Initial studies emphasized governance, and fiscal policy; later work expanded to government debt, public debt, and financial crisis and the most recent research highlights economic growth, local government debt, fiscal rules, and fixed effects.

Practical Implications—The findings offer several actionable implications for policymakers, managers, and stakeholders involved in debt management and governance. For debt management offices, the results highlight the need to move toward capability-driven operating models. This includes investing in specialized talent, continuous training, and interdisciplinary teams capable of handling complex tasks such as sovereign debt restructuring, fiscal risk analysis, and international negotiations. For HR professionals, the study emphasizes the importance of aligning HR practices with governance objectives. Recruitment, training, and performance management systems should be designed to build strategic competencies rather than merely support administrative functions. For policymakers, the findings underline the necessity of establishing clear frameworks and standards for HR governance in public financial institutions. This includes promoting transparency, accountability, and competency-based management aligned with international standards such as ISO 30408. Finally, for the broader research and policy community, the study suggests that strengthening HR governance in debt offices can improve not only organizational performance but also macroeconomic stability.

Limitation—This study is subject to certain limitations. First, the exclusive reliance on the Scopus database alone instead of mixing the results from different databases, while comprehensive focused on the Economics, Econometrics, Finance, and Business, Management, Accounting subject area. Second, the limitations stem from the scarcity of literature explicitly examining HR and governance interactions within debt offices, necessitating the inclusion of broader, related terms to construct a meaningful dataset. Third the choice of software and the specific parameters used for the network analysis can be another limitation. Fourth, the focus on peer-reviewed journal articles and reviews, while ensuring high academic rigor, excludes potentially valuable insights from books, book chapters, conference proceedings, and preprints. Incorporating these sources in future studies could provide a more comprehensive perspective and capture emerging trends (Islami & Mulolli, 2024). Overall recognizing these limitations, this study synthesized the current developments and emerging in the HR and Governance within Debt Offices domain, hopefully offering valuable insights and contributions to theory and practice.

References

Acemoglu, D., Johnson, S., & Robinson, J. A. (2001). The colonial origins of comparative development: An empirical investigation. *American Economic Review, 91*(5), 1369–1401. https://doi.org/10.1257/aer.91.5.1369

Alesina, A., & Tabellini, G. (1990). A positive theory of fiscal deficits and government debt. *The Review of Economic Studies, 57*(3), 403–414. https://doi.org/10.2307/2298021

Arellano, C. (2008). Default risk and income fluctuations in emerging economies. *American Economic Review, 98*(3), 690–712. https://doi.org/10.1257/aer.98.3.690

Aria, M., & Cuccurullo, C. (2017). Bibliometrix: An R-tool for comprehensive science mapping analysis. *Journal of Informetrics, 11*(4), 959–975. https://doi.org/10.1016/j.joi.2017.08.007

Armstrong, M. (2008). *Strategic human resource management: A guide to action* (4th ed.). Kogan Page.

Baas, J., Schotten, M., Plume, A., Côté, G., & Karimi, R. (2020). Scopus as a curated, high-quality bibliometric data source for academic research in quantitative science studies. *Quantitative Science Studies, 1*(1), 377–386. https://doi.org/10.1162/qss_a_00019

Ball, I. (2012). New development: Transparency in the public sector. *Public Money & Management, 32*(1), 35–40. https://doi.org/10.1080/09540962.2012.643054

Battaglini, M., & Coate, S. (2008). A dynamic theory of public spending, taxation, and debt. *American Economic Review, 98*(1), 201–236. https://doi.org/10.1257/aer.98.1.201

Ben Ali, T., & Ben Abdul Aziz Al Yahya, B. (2019). The effect of governance on public debt: An empirical investigation for the Arabian Gulf countries. *Journal of Economic Studies, 46*(4), 812–841. https://doi.org/10.1108/JES-07-2017-0168

Borresen, P., Cosio-Pascal, E., & Conférence des Nations Unies sur le commerce et le développement (Genève). (2002). *Role and organization of a debt office*. UN.

Caputo, A., Kargina, M., & Pellegrini, M. M. (2023). Conflict in virtual teams: A bibliometric analysis, systematic review, and research agenda. *International Journal of Conflict Management, 34*(1), 1–31. https://doi.org/10.1108/IJCMA-07-2021-0117

Caputo, A., Marzi, G., Maley, J., & Silic, M. (2019). Ten years of conflict management research 2007–2017: An update on themes, concepts and relationships. *International Journal of Conflict Management, 30*(1), 87–110. https://doi.org/10.1108/IJCMA-06-2018-0078

Caputo, A., Pizzi, S., Pellegrini, M. M., & Dabić, M. (2021). Digitalization and business models: Where are we going? A science map of the field. *Journal of Business Research, 123*, 489–501. https://doi.org/10.1016/j.jbusres.2020.09.053

Chen, Z., He, Z., & Liu, C. (2020). The financing of local government in China: Stimulus loan wanes and shadow banking waxes. *Journal of Financial Economics, 137*(1), 42–71. https://doi.org/10.1016/j.jfineco.2019.07.009

Coles, T., Dinan, C., & Hutchison, F. (2012). May we live in less interesting times? Changing public sector support for tourism in England during the sovereign debt crisis. *Journal of Destination Marketing & Management, 1*(1–2), 4–7. https://doi.org/10.1016/j.jdmm.2012.05.004

Dabic, M., González-Loureiro, M., & Harvey, M. (2015). Evolving research on expatriates: What is 'known'after four decades (1970–2012). *The International Journal of Human Resource Management, 26*(3), 316–337. https://doi.org/10.1080/09585192.2013.845238

Featherstone, K. (2011). The JCMS annual lecture: The Greek sovereign debt crisis and EMU: A failing state in a skewed regime. *JCMS: Journal of Common Market Studies, 49*(2), 193–217. https://doi.org/10.1111/j.1468-5965.2010.02139.x

Ferasso, M., Beliaeva, T., Kraus, S., Jones, P., & Gössling, T. (2025). Family business ethics: A literature review and research agenda. *Journal of Business Ethics,* 1–19. https://doi.org/10.1007/s10551-025-05951-9

Fukuyama, F. (2016). Governance: What do we know, and how do we know it? *Annual Review of Political Science, 19*(1), 89–105. https://doi.org/10.1146/annurev-polisci-042214-0442403

Howlett, M., & Ramesh, M. (2014). The two orders of governance failure: Design mismatches and policy capacity issues in modern governance. *Policy and Society, 33*(4), 317–327. https://doi.org/10.1016/j.polsoc.2014.10.002

Islami, X., & Mulolli, E. (2021). Linking HRM practices to operational performance in the emerging economy. *FEB Zagreb 12th international Odyssey conference on economics and business* (Vol. 3(1), pp. 891–901). Zagreb, Croatia.

Islami, X., & Mulolli, E. (2024). Human-artificial intelligence in management functions: A synergistic symbiosis relationship. *Applied Artificial Intelligence, 38*(1), 2439615. https://doi.org/10.1080/08839514.2024.2439615

ISO International Organization for Standardization. (2016). *Human resource management—Guidelines on human governance* (Reference number ISO 30408:2016(E)). Retrieved 2 March, 2026, from https://www.iso.org/obp/ui/#iso:std:iso:30408:ed-1:v1:en

Jobstreibizer, J., Beliaeva, T., Ferasso, M., Kraus, S., & Kallmuenzer, A. (2025). The impact of artificial intelligence on business models: A bibliometric-systematic literature review. *Management Decision, 63*(13), 372–396. https://doi.org/10.1108/MD-10-2024-2309

Kaehler, B., & Grundei, J. (2018). *HR governance: A theoretical introduction.* Springer.

Kessler, M. M. (1963). Bibliographic coupling between scientific papers. *American Documentation, 14*(1), 10–25. https://doi.org/10.1002/asi.5090140103

Kraus, S., Bouncken, R. B., & Yela Aránega, A. (2024). The burgeoning role of literature review articles in management research: An introduction and outlook. *Review of Managerial Science,* 18(2), 299–314. https://doi.org/10.1007/s11846-024-00729-1

Lewis, J. D., & Viñals, J. (2014). Revised guidelines for public debt management. *International Monetary Fund: Washington, DC, USA* (pp. 1–42). https://www.worldbank.org/content/dam/Worldbank/document/Debt/Revised%20Guidelines%20for%20Public%20Debt%20Management%202014_v2.pdf

Li, S., & Qi, T. (2023). Fiscal decentralization, local government debt, and corporate innovation. *Finance Research Letters, 54,* Article 103764. https://doi.org/10.1016/j.frl.2023.103764

Magnusson, T. (1999). *Legal arrangements for a debt office.* Swedish National Debt Office.

Mehmood, W., Mohd-Rashid, R., Khalid, A., Aman-Ullah, A., & Abbas, Y. A. (2022). Empirical analysis of country-level institutional quality and public debt: Perspective of South Asian countries. *Journal of Economic Cooperation & Development, 43*(1), 1–18.

Mercer Inc. Human Resource Consulting. (2003). Why HR governance matters—Managing the HR function for superior performance. Retrieved April 05, 2026, from https://www.adirelab.es/wp-content/uploads/2016/04/COMPLIANCE-WhyHRGovernanceMatters.pdf

Mulolli, E., Islami, X., & Mustafa, N. (2025). Individual vs. bundle HRM practices: effects on SMEs' performance. *Management: Journal of Contemporary Management Issues, 30*(2), 47–70. https://doi.org/10.30924/mjcmi.30.2.4

Page, M. J., McKenzie, J. E., Bossuyt, P. M., Boutron, I., Hoffmann, T. C., Mulrow, C. D., Moher, D., et al. (2021). The PRISMA 2020 statement: an updated guideline for reporting systematic reviews. *BMJ, 372*. https://doi.org/10.1136/bmj.n71

Tarek, B. A., & Ahmed, Z. (2017a). Governance and public debt accumulation: Quantitative analysis in MENA countries. *Economic Analysis and Policy, 56*, 1–13. https://doi.org/10.1016/j.eap.2017.06.004

Tarek, B. A., & Ahmed, Z. (2017b). Institutional quality and public debt accumulation: An empirical analysis. *International Economic Journal, 31*(3), 415–435. https://doi.org/10.1080/10168737.2017.1354906

Van Eck, N., & Waltman, L. (2023). VOSviewer manual. Manual for VOSviewer version 1.6.20, Universiteit Leiden. (n.d.). VOSviewer Manual. https://www.vosviewer.com/documentation/Manual_VOSviewer_1.6.20.pdf

Van Eck, N. J., & Waltman, L. (2016). VosViewer Manual: Manual for VosViewer Version 1.6. 5. CWTS, Leiden.

World Bank. (2021). *Debt management performance assessment methodology 2021 Edition*. World Bank. Retrieved April 2, 2026, from https://documents1.worldbank.org/curated/en/526391628746190611/pdf/Debt-Management-Performance-Assessment-Methodology-2021-Edition.pdf?

Yusuf, A., & Mohd, S. (2024). Investigating the asymmetric impact of public debt on economic growth in Nigeria. *Journal of the Knowledge Economy, 15*(2), 9452–9481. https://doi.org/10.1007/s13132-023-01362-1

Risk and Data: Physical/Transition Risk in Fiscal Scenarios and MTFF/MTEF (Medium-Term Expenditure Framework/ Medium-Term Fiscal Framework)

Roberta Bajrami, Kaltrina Bajraktari, Besime Ziberi, and Xhavit Islami

1 Introduction

Climate change generates material fiscal exposure for sovereign governments. Extreme weather events damage public infrastructure, reduce productivity, and disrupt economic activity, thereby diminishing tax revenues while simultaneously increasing public expenditure for disaster response, reconstruction, and adaptation. The transition toward low-carbon economic structures further alters investment patterns, energy markets, and public revenue streams linked to fossil fuel production. Empirical research identifies climate change as a macro-financial risk with direct implications for public finance and sovereign balance sheets (Agarwala et al., 2021; Feyen et al., 2020).

Climate-related shocks influence fiscal systems through both direct and indirect channels. Extreme weather events interrupt production, reduce agricultural output, and damage transport and energy infrastructure, prompting emergency spending and long-term adaptation commitments. Decarbonization policies alter fiscal capacity through changes in industrial activity, carbon taxation, and energy market restructuring (Agarwala et al., 2021; D'Orazio, 2025). Two categories of climate risk shape sovereign finance. Physical risks, floods, droughts, storms, and heatwaves, damage

R. Bajrami · K. Bajraktari · B. Ziberi · X. Islami (✉)
Faculty of Economics, AAB College, Pristina, Kosovo
e-mail: xhavit.islami@aab-edu.net

R. Bajrami
e-mail: roberta.bajrami@aab-edu.net

K. Bajraktari
e-mail: kaltrina.bajraktari@aab-edu.net

B. Ziberi
e-mail: besime.ziberi@aab-edu.net

H. Jarrar et al. (eds.), *Sovereign Sustainable Finance*, Sustainable Finance,
https://doi.org/10.1007/978-3-032-33239-4_9

infrastructure, reduce labour productivity, and generate significant macroeconomic losses in exposed economies (Bressan et al., 2024; Jin et al., 2025). Transition risks emerge from carbon pricing, environmental regulation, and technological substitution, which alter the profitability of carbon-intensive sectors and influence energy taxation and resource-related revenues (Daumas, 2023; Shen et al., 2025). Both categories affect sovereign balance sheets: climate-related disasters widen fiscal deficits, contribute to rising public debt ratios, and influence sovereign bond pricing. Evidence from European financial markets confirms that climate risk exposure affects asset valuation and sovereign risk premia (Bartolini et al., 2025), while macro-financial transmission studies demonstrate propagation through credit markets and investment portfolios (Battiston et al., 2020; Feyen et al., 2020).

Despite these developments, many fiscal planning systems treat climate shocks as irregular disturbances rather than structural fiscal drivers. Medium-term fiscal and expenditure frameworks typically rely on deterministic macroeconomic assumptions, rarely incorporating climate-related uncertainty, thereby underestimating long-term environmental and transition-related fiscal exposure (Raudla et al., 2022; Valencia & Martínez, 2025). Recent policy research advocates integrating climate scenarios, fiscal stress testing, and climate-informed macroeconomic modelling within fiscal planning to support forward-looking risk management (Allen et al., 2020; Oura et al., 2022). Such integration additionally supports sovereign sustainable finance instruments, green bonds and sustainability-linked debt, where transparent climate risk assessment strengthens investor confidence and fiscal credibility (Gupta et al., 2023). Climate-informed fiscal planning thereby improves transparency, strengthens risk management, and aligns fiscal policy with mitigation and adaptation objectives, reducing systemic vulnerability to climate shocks (Campiglio & Van der Ploeg, 2022; D'Orazio, 2025).

Recent advances in climate-aware fiscal policy have established climate change as a macro-financial risk with direct implications for sovereign balance sheets (Agarwala et al., 2021; Feyen et al., 2020), and there is growing integration of climate scenarios into financial sector assessments (Allen et al., 2020; Oura et al., 2022). Policy frameworks increasingly recognize the need for climate-informed fiscal planning to support sovereign sustainable finance instruments such as green bonds and sustainability-linked debt (Gupta et al., 2023). However, despite these developments, significant challenges remain. MTFF in emerging economies continue to treat climate shocks as irregular disturbances rather than structural fiscal drivers (Raudla et al., 2022; Valencia & Martínez, 2025). Practical methodologies for embedding physical and transition risks into MTFF and MTEF remain underdeveloped, particularly in data-constrained environments. The operational link between climate risk assessment and sovereign debt management strategies requires further elaboration, and systematic guidance for policymakers operating under institutional and capacity constraints is limited.

This chapter addresses these gaps by developing a structured, operational framework for integrating climate risks into fiscal planning in emerging economies. It contributes to the literature by: (i) systematically mapping transmission channels of physical and transition risks to fiscal variables within MTFF and MTEF, (ii) demonstrating practical integration methodologies through stress testing and sensitivity analysis, and (iii) bridging climate economics, sovereign risk analysis, and public financial management to provide actionable guidance for policymakers operating under capacity and data constraints. By synthesizing existing academic and policy literature and proposing an operational framework, this chapter enriches both theoretical understanding and practical application of climate-informed fiscal governance.

The chapter is organized as follows: Sect. 1 examines the relationship between climate risk and sovereign fiscal sustainability; Sect. 2 reviews fiscal transmission channels of climate risks; Sect. 3 evaluates approaches for integrating climate risks into fiscal frameworks and sovereign financing strategies; Sect. 4 discusses data and modelling challenges; and the conclusion presents policy implications for strengthening fiscal governance under climate uncertainty.

2 Climate Risk and Sovereign Fiscal Sustainability

Climate change functions as a structural determinant of macroeconomic stability and sovereign fiscal performance. Research identifies climate-related events as systemic risks with implications for fiscal stability, financial markets, and sovereign debt sustainability. Exposure affects fiscal balances through disruptions to economic activity, increased public expenditure, and altered revenue structures, linking environmental shocks with sovereign balance sheets and long-term fiscal stability (Battiston et al., 2020; Feyen et al., 2020).

Fiscal sustainability denotes the ability of governments to maintain stable public finances over time without persistent debt growth or disruptive adjustment. Climate change increases uncertainty in fiscal projections by affecting growth, tax revenues, expenditures, and financial stability simultaneously. Frameworks ignoring climate risk underestimate fiscal vulnerability, making climate risk assessment central to modern fiscal policy and sovereign risk analysis (Agarwala et al., 2021; Valencia & Martínez, 2025). Two main forms of climate-related risk are physical and transition risks. Physical risks stem from environmental hazards, while transition risks arise from economic restructuring linked to decarbonization policies, technological innovation, and energy system transformation. Both influence fiscal sustainability through distinct but interacting channels (Daumas, 2023; Di Febo & Angelini, 2025).

Physical risks include extreme weather events and slow-onset processes such as sea level rise or desertification, causing infrastructure damage, reduced productivity, and public health challenges (Bressan et al., 2024; Jin et al., 2025). Fiscal consequences include emergency spending for disaster response, reconstruction, and support to households and businesses. Adaptation investments in climate-resilient infrastructure further increase fiscal pressure, particularly in countries with limited resilience (Hochrainer-Stigler et al., 2024). Macroeconomic channels amplify these effects as output reductions decrease tax revenues. Empirical studies confirm that climate disasters increase fiscal deficits and debt ratios (Agarwala et al., 2021;

Gambhir et al., 2021). Selected estimates of climate-related economic losses reported in recent studies are presented in Table 1.

Transition risks arise from adjustments toward low-carbon systems. Policy instruments include carbon pricing, emissions trading, renewable energy subsidies, and regulation of carbon-intensive sectors. These reshape energy markets, industrial structures, and fiscal revenue systems (Campiglio & Van Der Ploeg, 2022). Carbon pricing increases production costs and reduces fossil fuel demand, affecting energy taxes and royalties. Resource-dependent economies face reduced fiscal revenues, requiring diversification and fiscal redesign (D'Orazio, 2025). Labor displacement in high-emission sectors increases social protection expenditures. Technological innovation alters asset profitability, creating stranded public and private energy assets (Kacperczyk, 2025; Semieniuk et al., 2020). Key fiscal implications of climate transition risk are summarized in Table 2.

Table 1 Selected estimates of economic losses from climate events

Study	Region	Climate event	Estimated economic impact
Agarwala et al. (2021)	Global sample	Climate disasters	Fiscal costs exceed 1 to 2% of GDP in affected economies
Bressan et al. (2024)	Global asset level analysis	Physical climate risk	Infrastructure exposure increases adaptation finance demand significantly
Estrada et al. (2024)	Mexico	Physical climate risk scenarios	Climate events generate substantial financial sector losses under severe warming pathways
Gambhir et al. (2021)	Global modelling	Climate transition and physical risk pathways	Economic damages increase under high emission scenarios

Source First author synthesis based on cited studies

Table 2 Fiscal effects of climate transition policies

Transition driver	Economic impact	Fiscal implication
Carbon pricing policies	Higher production costs in carbon intensive sectors	Reduced fossil fuel demand and changes in energy tax revenues
Renewable energy adoption	Energy market restructuring	Declining fossil fuel royalties
Technological innovation	Replacement of carbon intensive technologies	Stranded public and private energy assets
Industrial restructuring	Labor displacement in high emission sectors	Higher expenditure on labor market adjustment

Source First author synthesis based on cited studies

Table 3 Climate risks and fiscal sustainability

Climate risk	Economic effects	Fiscal consequences
Extreme weather events	Infrastructure damage and productivity loss	Higher reconstruction spending
Sea level rise	Coastal infrastructure losses	Increased adaptation investment
Decarbonisation policies	Structural industrial adjustment	Changes in energy taxation and fiscal revenues
Technological change	Energy market restructuring	Fiscal policy redesign and stranded asset exposure

Source First author synthesis based on cited studies

Physical and transition risks interact, producing complex fiscal dynamics. Climate disasters reduce output and revenues while increasing expenditure, generating persistent deficits and rising debt ratios. Dynamic macroeconomic models show repeated climate shocks raise public debt, particularly in economies with limited adaptation capacity (Gao, 2024). Financial markets incorporate climate risk into sovereign credit assessments, increasing bond risk premiums for vulnerable countries (Bartolini et al., 2025).

Governments act as insurers of last resort following disasters, creating contingent liabilities. Public infrastructure faces direct exposure, and damage reduces capital stock value while increasing reconstruction costs (Allen et al., 2020). Fiscal resilience, defined as the capacity to absorb shocks without undermining stability, requires systematic climate risk assessment, effective disaster management, and investment in resilient infrastructure (Hochrainer-Stigler et al., 2024).

Climate-informed fiscal frameworks integrate climate risk into expenditure planning and debt sustainability analysis. MTFF and MTEF provide mechanisms to project fiscal variables under environmental uncertainty (Raudla et al., 2022; Valencia & Martínez, 2025). Tools include climate stress testing, budget tagging, and fiscal risk statements to enhance transparency. Table 3 presents a conceptual overview of the relationship between climate risks and fiscal sustainability.

Climate change represents an expanding source of fiscal risk. Physical risks drive disaster-related spending and economic disruption, while transition risks reshape revenues and industrial structures. Integrating climate risk into fiscal policy is essential to maintain fiscal sustainability under rising environmental uncertainty (Campiglio & Van Der Ploeg, 2022; Feyen et al., 2020).

3 Fiscal Transmission Channels of Climate Risks

Climate change affects public finance through three main fiscal transmission channels: revenue, expenditure, and contingent liabilities. These channels link environmental shocks and climate transition processes with fiscal outcomes, including deficits, debt accumulation, and borrowing conditions (Agarwala et al., 2021; D'Orazio, 2025). Climate shocks produce asymmetric fiscal responses, with immediate expenditure increases and slower revenue recovery, creating fiscal pressure and higher borrowing requirements (Daumas, 2023; Shen et al., 2025). Understanding these channels supports climate risk integration into fiscal policy, debt sustainability analysis, and MTFF and MTEF.

The revenue channel reflects the impact of climate change on economic output and taxable income. Disruptions reduce production, employment, and trade activity, weakening government revenue bases. Agriculture is highly climate-sensitive. Temperature changes, floods, and droughts reduce crop yields and livestock productivity. In many developing economies, agriculture contributes 20–30% of GDP and employment, making shocks highly influential on income, consumption, and export taxes (Agarwala et al., 2021). Industrial production also suffers from climate events. Floods, storms, and heatwaves damage factories, reduce labor productivity, and interrupt supply chains. Infrastructure damage affects transport networks and electricity supply, lowering corporate tax receipts. Climate-related disruptions to trade reduce customs duties and export revenues. Decarbonization policies, carbon pricing, and renewable energy adoption further alter energy consumption patterns and industrial structures. Resource-dependent economies experience declining energy-related revenues, while technological advances in energy efficiency and renewable expansion reduce fossil fuel taxation, though carbon pricing can offset some losses (D'Orazio, 2025; Daumas, 2023). Table 4 summarizes the impact of climate disasters on fiscal revenue.

The expenditure channel reflects direct fiscal costs of climate-related disasters. Emergency response involves disaster relief, evacuation, healthcare, and humanitarian assistance. Reconstruction spending restores infrastructure and public services. Long-term adaptation investment supports climate-resilient infrastructure, including flood barriers, irrigation, and urban flood management (Agarwala et al., 2021; Hochrainer-Stigler et al., 2024). Climate-related health and social protection costs further increase expenditure obligations. Table 5 presents major components of climate-related public expenditure.

Climate risks create contingent liabilities, potential obligations triggered by specific events. Governments often act as insurers of last resort, providing financial support to households, businesses, and public enterprises after disasters. Losses

Table 4 Impact of climate disasters on fiscal revenue

Sector	Climate impact	Fiscal effect
Agriculture	Crop failure, livestock losses	Lower income tax and export revenues
Industry	Infrastructure damage, production disruption	Reduced corporate tax revenues
Trade	Supply chain disruption	Lower customs and trade taxes

Source Adapted from Agarwala et al. (2021) and Shen et al. (2025)

Table 5 Major components of climate-related public expenditure

Expenditure category	Fiscal activity	Examples
Emergency response	Immediate disaster relief	Evacuation operations, medical services
Reconstruction	Infrastructure restoration	Roads, bridges, power systems
Adaptation investment	Long term resilience measures	Flood barriers, irrigation systems

Source Synthesized from Hochrainer-Stigler et al. (2024) and Agarwala et al. (2021)

affecting infrastructure, utilities, and adaptation projects activate government guarantees, increasing fiscal exposure. Climate shocks can also threaten financial sector stability, prompting public interventions through guarantees, liquidity support, or recapitalization (Shen et al., 2025). Table 6 presents fiscal transmission mechanisms of climate risk.

Revenue reductions, expenditure increases, and contingent liabilities combine to widen fiscal deficits and raise public debt. Borrowing to finance reconstruction and adaptation further amplifies debt ratios. Financial markets assess sovereign climate vulnerability and adaptation capacity, pricing higher risk into sovereign bonds. Climate-exposed countries therefore face elevated borrowing costs, higher risk premiums, and structural fiscal pressure (Bartolini et al., 2025). Integrating these transmission mechanisms into fiscal frameworks enhances government capacity to manage climate-related fiscal exposure and maintain long-term fiscal stability. Figure 1 illustrates the relationship between climate risk transmission and sovereign fiscal outcomes.

Table 6 Fiscal transmission mechanisms of climate risk

Transmission channel	Mechanism	Fiscal impact
Revenue channel	Reduced economic activity	Lower tax revenues
Expenditure channel	Disaster response and adaptation investment	Higher public spending
Contingent liabilities	Government guarantees and financial support	Increased fiscal risk exposure

Source Adapted from Daumas (2023) and Shen et al. (2025)

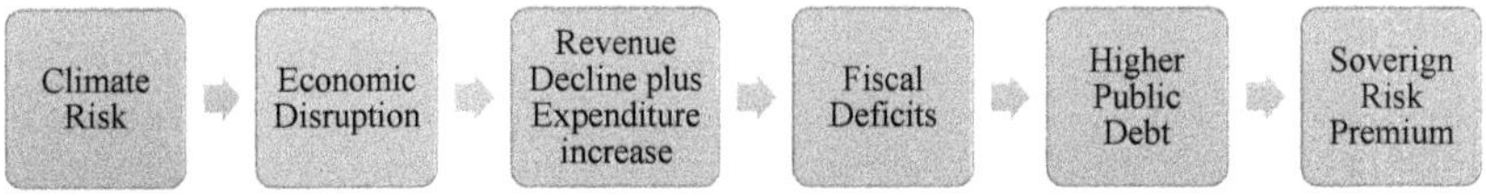

Fig. 1 Climate risk transmission and sovereign fiscal outcomes

4 Approaches for Integrating Climate Risks into Fiscal Frameworks and Sovereign Financing Strategies

MTFF and MTEF are central instruments in public financial management. They structure fiscal planning across three-to-five-year horizons, projecting revenue performance, expenditure ceilings, fiscal balances, and debt trajectories. MTFF focuses on macro fiscal sustainability, projecting aggregate revenue, expenditure, deficits, and debt. MTEF complements this through sector-level expenditure planning, linking fiscal ceilings to public policy priorities and aligning annual budgets with long-term strategies. Conventional frameworks rely on deterministic forecasts that assume stable economic conditions, omitting climate risks. This omission increases fiscal vulnerability as climate shocks generate unexpected expenditure pressures and persistent revenue losses (Agarwala et al., 2021; D'Orazio, 2025).

Scenario analysis is a primary method for integrating climate risks into fiscal frameworks. It evaluates fiscal outcomes under alternative climate and economic conditions. Policymakers construct trajectories based on varying climate pathways, including natural disaster frequency, sea level rise, and structural adjustments from energy transitions. Scenario analysis reveals vulnerabilities in revenue and expenditure, supporting contingency planning (Valencia & D'Orazio, 2025; Martínez, 2025). Table 7 presents climate scenario pathways in fiscal projections.

Fiscal stress testing simulates adverse conditions to assess fiscal resilience. Events include large-scale floods, prolonged droughts, or abrupt decarbonization. Stress tests identify vulnerabilities in revenue systems and expenditure commitments, informing fiscal buffers, contingency funds, disaster insurance, and precautionary borrowing strategies. Stress testing also strengthens transparency by integrating fiscal risk statements with annual budgets (Agarwala et al., 2021; Gao, 2024). Table 8 presents fiscal stress testing indicators for climate risk.

Disaster risk models estimate expected economic and fiscal costs of natural hazards using historical data, climate projections, and exposure indicators. These models calculate expected annual losses and potential fiscal obligations, informing

Table 7 Climate scenario pathways in fiscal projections

Scenario type	Climate assumptions	Fiscal implications
Moderate climate pathway	Gradual temperature increases and limited disaster intensity	Moderate decline in agricultural output and small fiscal expenditure increases
Severe climate pathway	Frequent floods, droughts, and heatwaves	Significant reconstruction costs and reduced tax revenues
Accelerated energy transition	Rapid decarbonization and reduced fossil fuel demand	Decline in fossil fuel taxation and restructuring of energy subsidies

Source Adapted from Valencia and Martínez (2025) and D'Orazio (2025)

Table 8 Fiscal stress testing indicators for climate risk

Indicator	Description	Fiscal relevance
GDP shock	Decline in economic output following climate disaster	Reduces tax revenues
Infrastructure loss	Damage to public capital stock	Increases reconstruction spending
Agricultural output decline	Reduction in crop productivity	Weakens export earnings and rural income taxes
Energy transition shock	Reduction in fossil fuel demand	Decline in fuel taxation revenues

Source Agarwala et al. (2021) and Gao (2024)

contingency reserves. Empirical evidence shows average annual disaster losses of 1–2 percent of GDP in climate-vulnerable economies, with extreme events exceeding 5% (Gao, 2024). Table 9 presents estimated fiscal costs of climate disasters.

Climate variables integrated into macroeconomic models produce more realistic projections of growth, sectoral output, and fiscal requirements. These models capture interactions between environmental change, productivity, and fiscal performance, providing long-term projections for revenues and expenditures. Figure 2 illustrates the climate-adjusted fiscal forecasting mechanism.

Table 9 Estimated fiscal costs of climate disasters

Region	Average annual disaster loss as percent of GDP	Major climate hazard
Small island economies	2.5	Tropical storms and coastal flooding
Sub-Saharan Africa	1.2	Droughts and agricultural shocks
South Asia	1.8	Floods and cyclones
Southern Europe	0.6	Heatwaves and wildfires

Source First author synthesis based on Gao (2024) and Agarwala et al. (2021)

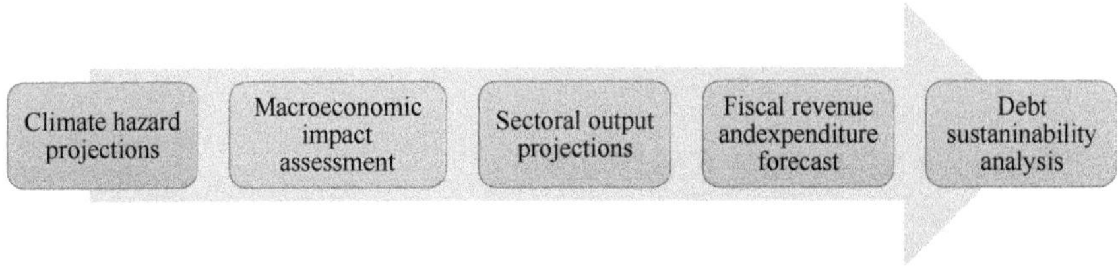

Fig. 2 Climate-adjusted fiscal forecasting mechanism

Table 10 Growth of sovereign green bond issuance

Year	Global issuance value (USD billions)
2018	21
2020	45
2022	135
2023	210

Source D'Orazio (2025)

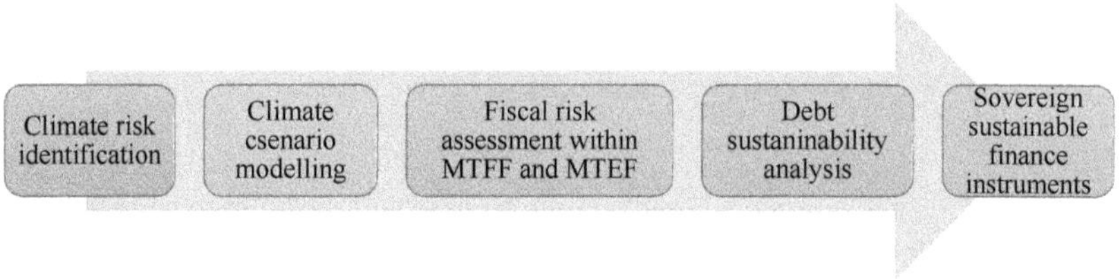

Fig. 3 Climate risk integration within fiscal policy

Climate-informed frameworks support sovereign sustainable finance, including green and sustainability-linked bonds. Green bonds finance projects such as renewable energy, resilient transport, and flood protection. Sustainability-linked bonds tie interest costs to environmental target performance. Transparent climate-adjusted fiscal planning strengthens investor confidence and sovereign credit credibility. Global sovereign green bond issuance exceeded 210 billion USD in 2023, with emerging market participation increasing (D'Orazio, 2025). Climate budget tagging tracks mitigation and adaptation spending, improving monitoring and policy alignment. Institutional coordination among finance ministries, environmental agencies, and debt offices enhances alignment between fiscal governance, climate objectives, and debt management. Table 10 presents the growth of sovereign green bond issuance.

Climate risk integration follows a structured process linking risk assessment, scenario analysis, and fiscal planning. Scenario outputs inform medium-term planning, while debt sustainability analysis evaluates impacts on borrowing capacity. Integration strengthens fiscal resilience, transparency, and government capacity to manage climate-related fiscal shocks and maintain long-term stability. Figure 3 illustrates climate risk integration within fiscal policy.

5 Data and Modelling Challenges

Integration of climate risk into fiscal planning frameworks faces significant empirical and methodological constraints. Recognition of climate-related fiscal risk has increased, but practical implementation remains limited. Major constraints arise from limited data availability, modelling uncertainty, and difficulty linking environmental processes with macroeconomic and fiscal systems. These affect integration of climate risk into MTFF, MTEF, and sovereign debt sustainability analysis (Agarwala et al., 2021; Gao, 2024).

Reliable climate-related fiscal data are often unavailable or fragmented. Meteorological agencies collect environmental indicators, disaster management agencies record damage estimates, infrastructure ministries maintain asset inventories, and fiscal authorities maintain expenditure records. Integration across these datasets is rare, limiting the estimation of fiscal consequences of climate events. Emerging economies face more severe data gaps, with incomplete historical records of disaster-related expenditure (Agarwala et al., 2021). Satellite-based environmental monitoring provides consistent data on temperature trends, precipitation, vegetation, and sea level variation. Global disaster loss databases, such as EM-DAT, compile information on disaster frequency, economic losses, and human impacts, providing baseline data for fiscal risk modelling. Table 11 presents key climate risk data sources for fiscal analysis.

Climate scenario modelling faces methodological challenges. Climate change evolves over long horizons, while MTFF operate over three to five years. Bridging these horizons requires assumptions linking long-term climate pathways to short-term fiscal impacts. Scenario frameworks, such as NGFS pathways, combine physical and transition risk projections with macroeconomic and financial system outcomes (Monasterolo et al., 2023). Table 12 presents climate scenario categories used in fiscal analysis.

Granular assessment of public assets improves fiscal exposure estimates. Infrastructure such as roads, bridges, ports, energy networks, and coastal assets are vulnerable to floods, heatwaves, and sea level rise. Asset-level risk analysis combines inventories with geospatial hazard maps to estimate probabilities and costs of damage, guiding adaptation investment (Bressan et al., 2024). Table 13 presents infrastructure exposure to climate hazards.

Climate risks affect sovereign borrowing and investor perceptions. Physical hazards, frequent disasters, and energy transition risks influence creditworthiness and borrowing costs. Countries with higher climate exposure face higher sovereign bond

Table 11 Key climate risk data sources for fiscal analysis

Data source	Information provided	Application in fiscal analysis
Meteorological datasets	Temperature, precipitation, climate variability	Climate hazard projections
EM-DAT disaster database	Disaster frequency and economic losses	Estimation of expected fiscal losses
Satellite observations	Environmental monitoring and hazard mapping	Identification of climate exposure
Infrastructure asset inventories	Location and condition of public assets	Infrastructure vulnerability assessment

Source Agarwala et al. (2021) and Gao (2024)

Table 12 Climate scenario categories used in fiscal analysis

Scenario type	Characteristics	Fiscal implications
Orderly transition	Gradual decarbonization and early climate policy implementation	Moderate fiscal adjustment and stable transition costs
Delayed transition	Late policy response and rapid emissions reduction after delay	High short term fiscal adjustment costs
Hot house world	Limited mitigation and severe climate impacts	Large disaster costs and persistent fiscal deficits

Source Monasterolo et al. (2023)

Table 13 Infrastructure exposure to climate hazards

Infrastructure type	Major climate hazard	Fiscal implication
Transport infrastructure	Flooding and extreme rainfall	Reconstruction and maintenance expenditure
Energy infrastructure	Heatwaves and storms	Power system repair costs
Water infrastructure	Drought and precipitation variability	Adaptation investment
Coastal infrastructure	Sea level rise and storm surges	Coastal protection expenditure

Source Bressan et al. (2024)

spreads. Adaptation capacity mitigates risk premiums, while recurring disasters and fossil fuel dependence increase fiscal vulnerability (Bartolini et al., 2025; D'Orazio, 2025). Table 14 presents climate risk and sovereign bond market indicators.

Integration of climate risk into fiscal planning requires reliable datasets, scenario frameworks, and interdisciplinary analytical capacity. Advances in satellite monitoring, disaster databases, scenario modelling, and asset-level analysis improve tools for climate-informed policy. Strengthened climate risk analytics enhance fiscal resilience, support medium-term planning, and facilitate sustainable sovereign finance strategies.

Table 14 Climate risk and sovereign bond market indicators

Indicator	Market interpretation	Fiscal implication
Climate vulnerability index	Exposure to physical climate hazards	Higher borrowing costs
Energy transition risk	Dependence on fossil fuel revenues	Revenue volatility
Disaster frequency	Recurring climate shocks	Increased fiscal deficits
Adaptation capacity	Institutional ability to manage climate risk	Lower sovereign risk premium

Source Bartolini et al. (2025) and D'Orazio (2025)

6 Conclusion

Climate change constitutes a structural fiscal risk for sovereign governments, distinct in character from the cyclical macroeconomic shocks that conventionally inform fiscal planning. Environmental processes, encompassing both the physical manifestations of a warming climate and the economic consequences of decarbonization policy, alter patterns of economic activity, reshape fiscal balances, and influence financial market perceptions over extended time horizons, necessitating a fundamental reorientation of fiscal risk management.

Theoretically, this chapter advances the understanding of climate change as a structural fiscal factor rather than an exogenous shock. By systematically mapping transmission channels through revenue, expenditure, and contingent liability mechanisms, it extends the macro-fiscal risk literature and bridges climate economics with public financial management theory. The framework demonstrates how physical and transition risks interact dynamically to influence sovereign balance sheets and debt sustainability trajectories. It conceptualizes fiscal resilience not merely as shock absorption capacity but as systematic integration of climate uncertainty into medium-term planning instruments. This theoretical contribution enriches academic discourse by positioning climate risk as a permanent structural variable in fiscal sustainability analysis, rather than a temporary disturbance.

Practically, the chapter provides actionable methodologies for fiscal policymakers in emerging economies. The integration of scenario analysis, stress testing, and disaster risk modeling into MTFF and MTEF offers operational tools that enhance forecasting accuracy, improve contingent liability identification, and strengthen fiscal resilience. The framework supports climate-informed sovereign financing strategies, facilitating access to green bonds and sustainability-linked instruments through transparent climate risk assessment. By demonstrating how climate variables can be embedded within existing fiscal planning architectures, the chapter provides practical guidance for ministries of finance, debt management offices, and budget authorities operating under data and capacity constraints. The structured approach enables governments to design fiscal policies coherently aligned with adaptation and mitigation priorities while maintaining debt sustainability.

Physical climate risks generate substantial and increasingly recurrent direct fiscal costs. Extreme weather events inflict damage upon public infrastructure, disrupt economic production, and diminish taxable output, while governments bear responsibility for funding emergency response, reconstruction, and long-term adaptation investments. These burdens fall disproportionately upon economies characterised by limited fiscal buffers or elevated environmental exposure, compounding pre-existing public finance vulnerabilities. Transition risks produce equally consequential structural changes, as decarbonisation policies fundamentally alter energy markets and industrial structures. For governments dependent upon hydrocarbon revenues, the progressive erosion of fossil fuel tax bases represents a particularly acute long-term fiscal challenge requiring comprehensive revenue restructuring and public investment redirection.

The macroeconomic transmission channels through which climate risks compromise fiscal sustainability are multiple and mutually reinforcing. Climate-induced output losses reduce tax revenues while simultaneously elevating public expenditure, generating persistent fiscal deficits and rising debt ratios that influence sovereign risk assessments and borrowing costs. Adverse feedback dynamics between environmental exposure, fiscal deterioration, and financial market discipline may consequently emerge. Notwithstanding these recognised risks, traditional fiscal planning frameworks, reliant upon deterministic macroeconomic assumptions, systematically underestimate long-term fiscal exposure by failing to incorporate climate hazard distributions, infrastructure vulnerability, or transition policy shocks.

Climate-informed fiscal planning offers a substantive corrective. Scenario analysis, fiscal stress testing, and probabilistic disaster risk modelling collectively enhance forecasting accuracy, improve contingent liability identification, and strengthen the evidential basis for policy preparation. The effectiveness of these frameworks depends upon improved climate data systems and the integration of climate science with macroeconomic modelling. Climate-informed frameworks additionally support sovereign sustainable finance strategies, as transparent climate risk integration within fiscal projections strengthens investor confidence and facilitates access to expanding sustainability-oriented capital markets.

Policy implications for strengthening fiscal governance under climate uncertainty include: (i) systematic integration of climate scenarios into MTFF and MTEF to replace deterministic assumptions with climate-adjusted projections, (ii) establishment of contingency reserves and fiscal buffers based on probabilistic disaster risk assessments, (iii) development of institutional coordination mechanisms linking finance ministries, environmental agencies, and debt management offices, (iv) implementation of climate budget tagging systems to track adaptation and mitigation expenditures, (v) strengthening of data infrastructure through satellite monitoring, disaster databases, and asset inventories, and (vi) alignment of sovereign financing strategies with climate-informed fiscal planning to access green and sustainability-linked capital markets.

In summation, recognizing climate change as a core macro-fiscal risk enables governments to design fiscal policies coherently aligned with adaptation and mitigation priorities. Strengthening climate integration within fiscal forecasting, budgetary procedures, and debt management is essential for sustaining sovereign fiscal resilience under conditions of accelerating environmental change and deepening energy transition. This chapter contributes to both theoretical advancement and practical application by providing a structured framework that bridges academic research, policy development, and operational implementation in the context of climate-aware fiscal governance.

References

Agarwala, M., Burke, M., Klusak, P., Mohaddes, K., Volz, U., & Zenghelis, D. (2021). Climate change and fiscal sustainability: Risks and opportunities. *National Institute Economic Review, 258*(1), 28–46. https://doi.org/10.1017/nie.2021.37

Allen, T., Dées, S., Graciano, C., Chouard, V., Clerc, L., De Gaye, A., Devulder, A., Diot, S., Lisack, N., Pegoraro, F., Rabaté, M., Svartzman, R., & Vernet, L. (2020). *Climate-related scenarios for financial stability assessment: An application to France* (Banque de France Working Paper 774). https://doi.org/10.2139/ssrn.3653131

Bartolini, N., Romagnoli, S., & Santini, A. (2025). Understanding climate risk in Europe: Are transition and physical risk priced in equity and fixed-income markets? *Journal of Empirical Finance, 81*, Article 101672. https://doi.org/10.1016/j.jempfin.2025.101672

Battiston, S., Dafermos, Y., & Monasterolo, I. (2020). Climate risks and financial stability. *Journal of Financial Stability, 54*, Article 100867. https://doi.org/10.1016/j.jfs.2021.100867

Bressan, G., Đuranović, A., Monasterolo, I., & Battiston, S. (2024). Asset-level assessment of climate physical risk matters for adaptation finance. *Nature Communications, 15*, 3776. https://doi.org/10.1038/s41467-024-48820-1

Campiglio, E., & Van Der Ploeg, F. (2022). Macrofinancial risks of the transition to a low-carbon economy. *Review of Environmental Economics and Policy, 16*(2), 173–195. https://doi.org/10.1086/721016

D'Orazio, P. (2025). Addressing climate risks through fiscal policy in emerging and developing economies: What do we know and what lies ahead? *Energy Research & Social Science, 119*, Article 103852. https://doi.org/10.1016/j.erss.2024.103852

Daumas, L. (2023). Financial stability, stranded assets and the low-carbon transition: A critical review of the theoretical and applied literatures. *Journal of Economic Surveys, 37*(5), 1560–1594. https://doi.org/10.1111/joes.12551

Di Febo, E., & Angelini, E. (2025). Transition risk in climate change: A literature review. *Risks, 13*(4), 66. https://doi.org/10.3390/risks13040066

Estrada, F., del Carmen, M. A. A., Calderon-Bustamante, O., Botzen, W. J., Martinez-Jaramillo, S., & Battiston, S. (2024). Assessing the physical risks of climate change for the financial sector: A case study from Mexico's Central Bank. *Climate Risk Management, 46*, Article 100634. https://doi.org/10.1016/j.crm.2024.100634

Feyen, E., Utz, R., Huertas, I., Bogdan, O., & Moon, J. (2020). *Macro-financial aspects of climate change* (World Bank Policy Research Working Paper, 9109). https://doi.org/10.1596/1813-9450-9109

Gambhir, A., George, M., Mcjeon, H., Arnell, N., Bernie, D., Mittal, S., Köberle, A., Lowe, J., Rogelj, J., & Monteith, S. (2021). Near-term transition and longer-term physical climate risks of greenhouse gas emissions pathways. *Nature Climate Change, 12*(1), 88–96. https://doi.org/10.1038/s41558-021-01236-x

Gao, S. (2024). An exogenous risk in fiscal-financial sustainability: Dynamic stochastic general equilibrium analysis of climate physical risk and adaptation cost. *Journal of Risk and Financial Management, 17*(6), 244. https://doi.org/10.3390/jrfm17060244

Gupta, B., Cheng, R., & Rajan, R. (2023). Climate risks: Nexus between green financial policies and fiscal space. *Applied Economics Letters, 31*(18), 1656–1660. https://doi.org/10.1080/13504851.2023.2205092

Hochrainer-Stigler, S., Trogrlić, R., & Reiter, K. (2024). Fiscal resilience over time and its management in the context of multi-risks: An application to the Danube Region. *Natural Hazards, 120*(13), 12163–12180. https://doi.org/10.1007/s11069-024-06678-y

Jin, Y., Zhu, H., Gu, X., & Bo, N. (2025). Climate risk and actual losses: Physical climate risks vs. transition climate risks. *Advances in Economics, Management and Political Sciences, 29*, 380. https://doi.org/10.54254/2754-1169/2025.bj29380

Kacperczyk, M. (2025). Transition risk in the banking sector. *Nature Climate Change, 15*(9), 923–924. https://doi.org/10.1038/s41558-025-02400-3

Monasterolo, I., Nieto, M., & Schets, E. (2023). *The good, the bad and the hot house world: Conceptual underpinnings of the NGFS scenarios and suggestions for improvement* (ECB Working Paper, 2803). https://doi.org/10.2139/ssrn.4211384

Oura, H., Krznar, I., Lipinsky, F., Gross, M., Grippa, P., Adrian, T., Lepore, C., Lamichhane, S., Panagiotopoulos, A., & Haksar, V. (2022). *Approaches to climate risk analysis in FSAPs* (IMF Staff Climate Notes, 2022/005). https://doi.org/10.5089/9798400212895.066

Raudla, R., Douglas, J., & MacCarthaigh, M. (2022). Medium-term expenditure frameworks: Credible instrument or mirage? *Public Budgeting & Finance, 42*(4), 3–29. https://doi.org/10.1111/pbaf.12316

Semieniuk, G., Campiglio, E., Mercure, J., Volz, U., & Edwards, N. (2020). Low-carbon transition risks for finance. *Wiley Interdisciplinary Reviews: Climate Change, 12*(1), Article e678. https://doi.org/10.1002/wcc.678

Shen, H., Feng, H., Xiao, Q., & Guo, H. (2025). Research on the dual effects of corporate physical and transition climate risks on total factor productivity. *Emerging Markets Finance and Trade, 61*(15), 4247–4265. https://doi.org/10.1080/1540496x.2025.2508880

Valencia, O., & Martínez, A. (2025). Enhancing fiscal resilience: Medium-term frameworks for managing emerging risks (*IDB Working Paper*, *1503*). https://doi.org/10.18235/0013503

Oversight and Assurance in Sovereign Sustainable Finance: SAIs, Internal Audit, and Performance Budgeting

Majdi Hassen

1 Introduction

Sovereign sustainable finance can lose credibility quickly when disclosure outruns control. A government may publish a strong framework, secure an external opinion, and attract a diversified investor base, yet still struggle to trace proceeds, verify climate tags, reconcile debt and budget information, or explain how financed activities link to stated outcomes. When that happens, the market problem and the public accountability problem become the same problem: credibility weakens because the control chain is incomplete (Di Marco et al., 2023; World Bank, 2022).

Assurance is therefore not a final layer added after issuance. It is part of the financing architecture. What matters is the chain that links bond frameworks, budget classifications, accounting records, debt reports, internal controls, audit work, and public disclosure. This chapter argues that strong oversight in sovereign sustainable finance requires coordination across three functions that are too often treated separately: internal control and internal audit inside government, supreme audit institutions outside the executive, and performance-informed budgeting systems that link financing to measurable public purposes.

Assurance has become central because sovereign sustainable finance now makes claims that cut across several administrative systems at once. A framework may say that proceeds finance eligible expenditures, that selection follows a defined methodology, that reporting will track results, and that budget information is consistent with bond disclosure. Each of those claims can be true in isolation and still fail as a package if the underlying systems do not reconcile. That is why assurance cannot be treated as a final review step added after issuance. It has to be designed into the financing chain from the beginning, when eligibility rules, reporting lines, and

M. Hassen (✉)
ESSEC Tunis, University of Tunis, Tunis, Tunisia
e-mail: majdi.hassen@gmail.com

H. Jarrar et al. (eds.), *Sovereign Sustainable Finance*, Sustainable Finance,
https://doi.org/10.1007/978-3-032-33239-4_10

evidence requirements are still being set (INTOSAI, 2019a, 2019b; World Bank, 2022).

The governance challenge is wider than simple fraud prevention. Category drift is a more common danger. Governments may tag expenditures as climate related without consistent criteria, publish allocation reports that do not map cleanly to budget execution, or rely on environmental estimates that are difficult to verify later. Di Marco et al. (2023) show in a broader sustainable-finance setting why transparency claims can quickly become fragile when the underlying information architecture is weak. In the sovereign context that fragility is amplified because legislatures, audit bodies, and citizens, not only investors, have a legitimate interest in whether the claims attached to borrowing are actually supportable (Di Marco et al., 2023; INTOSAI, 2019c).

2 Why Assurance Now Sits at the Center of Sovereign Sustainable Finance

Sustainable sovereign finance makes claims that go beyond ordinary borrowing. A green or sustainability-related instrument claims that proceeds or performance targets are tied to identifiable public purposes. A resilience clause claims that a contractual mechanism will operate fairly under stress. A guarantee-backed structure claims that public risk support is proportionate and monitored. Each of these claims can be priced by markets, but each also requires public verification. That is why assurance is central rather than peripheral (Bhutta et al., 2022; World Bank, 2022).

The problem is not merely fraud or misreporting. It is also category drift. Governments can easily build reporting systems that track transactions but not outcomes, or outcomes but not fiscal traceability. Di Marco et al. (2023) show in the broader sustainable finance context that "risk transparency" can become symbolic when disclosure is not backed by substantive governance. Sovereign finance faces the same danger. A framework that looks rigorous to investors may still sit on weak budget execution controls or uncertain project data.

Public accountability raises the standard further. Unlike corporate issuers, sovereigns operate within constitutional and budgetary systems that involve legislatures, audit bodies, and citizens, not only creditors. This means that external second-party opinions or verification providers cannot substitute for public audit and internal control. They play a useful market role, but they do not discharge the state's responsibility to show that public resources were allocated, managed, and reported according to law and policy intent (INTOSAI, 2019a, 2019b).

The practical implication is that oversight should begin early. Before issuance, governments need assurance over eligibility rules, data sources, governance arrangements, and reporting lines. During execution, they need traceability across budget and debt systems. After issuance, they need internal and external review that can test both compliance and effectiveness. Anything less leaves the sovereign exposed to greenwashing accusations, weak legislative scrutiny, and fragile market trust (Table 1).

A strong assurance architecture therefore begins before the first report is published. Governments need to decide who validates project eligibility, who owns the use-of-proceeds ledger, how budget tags relate to thematic frameworks, and what documentation must be retained for audit. These questions are operational, not ceremonial. Once issuance begins, weak role definition becomes expensive because the office then has to explain discrepancies after the fact. Public audit standards are useful here precisely because they emphasize criteria, evidence, and traceability. If the issuer cannot show how a figure in a public report is connected to an underlying record, assurance becomes largely rhetorical (INTOSAI, 2019a, 2019b; World Bank, 2022).

Green budgeting and budget tagging help only when they are methodologically disciplined. OECD work makes this clear. A tag can improve transparency, prioritization, and policy discussion, but only if it has explicit definitions, is applied consistently, and can be followed through execution and ex post reporting. Tags that are broad, unstable, or disconnected from accounting records create the appearance of control without the reality of it. PEFA's climate-responsive framework pushes in the same direction by asking whether climate considerations are embedded across core public financial management processes rather than attached as an afterthought (OECD, 2021, 2024; PEFA Secretariat, 2024).

Pre-issuance evidence packs can make a major difference to auditability later on. Before the bond is sold or the framework is published, the issuer can assemble the criteria, responsible units, data definitions, reconciliation rules, and retention expectations that will govern the life of the instrument. Doing that work early reduces the temptation to improvise after the market announcement has already created expectations. It also makes later assurance more efficient because reviewers can test a defined process rather than reconstructing one from scattered documents (The Institute of Internal Auditors, 2024; World Bank, 2022).

Reconciliation should be designed as a routine rather than an emergency exercise. Numbers published in debt reports, budget documents, allocation reports, and impact reports should be expected to match or to have explainable reasons for difference. When institutions only discover inconsistencies shortly before publication, the control problem is already well advanced. A stronger model uses periodic reconciliations and sign-off points so that traceability is maintained throughout the reporting cycle rather than rebuilt at the end of it (OECD, 2021; PEFA Secretariat, 2024).

3 Internal Controls, Budget Tagging, and Traceability

The first layer of assurance sits inside public financial management. Green budgeting and climate budget tagging are relevant here because they offer methods for identifying which policies, projects, or expenditures are considered climate-related at budget stage. OECD guidance shows that tagging can improve transparency and

Table 1 Assurance architecture for sovereign sustainable finance

Assurance actor	Main role	Typical questions asked	Evidence base
Management and first-line controls	Execute and document approved processes	Were funds allocated according to approved rules?	Budget records, debt records, project files
Internal audit	Independent risk-based assurance inside government	Do the controls work in design and operation?	Control testing, reconciliations, system logs, sample reviews
Supreme audit institution	Public-sector external audit and accountability	Were legality, regularity, performance, and value-for-money achieved?	Financial statements, performance evidence, audit fieldwork
External reviewer or second-party opinion provider	Market-oriented framework or disclosure review	Does the framework align with declared principles?	Framework documents, selected reporting and methodology statements
Legislature or parliamentary committee	Scrutinize disclosure, budget alignment, and public purpose	Are financing claims consistent with approved public spending and policy goals?	Budget documents, hearings, audit reports, allocation and impact reports
Scientific or technical reviewer	Test methodological assumptions where specialized evidence is needed	Are impact metrics and eligibility methods technically defensible?	Methodology notes, models, sector data, expert review memos
Public reporting and follow-up	Explain findings and corrective actions to stakeholders	Assurance results are published without operational response	Management responses, action plans, progress updates, and follow-up review
Corrective-action owner	Track remediation after assurance findings	Findings recur because no unit owns follow-up	Assigned responsibility, deadlines, progress logs, and escalation reporting
Evidence repository	Preserves audit trail across the financing cycle	Reviewers must reconstruct evidence late in the process	Centralized retention of frameworks, reconciliations, and supporting files

Source Author's own elaboration based on The Institute of Internal Auditors (2024), INTOSAI (2019a, 2019b), and World Bank (2022)

prioritization, but only if the tag is tied to a clear methodology and carried through execution and reporting. A tag that appears in budget preparation but disappears during implementation creates the illusion of control without providing traceability (OECD, 2021, 2024).

This is where climate-responsive PFM becomes operational. The PEFA supplementary framework on climate-responsive public financial management emphasizes that climate issues should be integrated into mainstream PFM processes rather than treated as detached annexes. For sustainable borrowing, that means the debt office's disclosure and the budget system's classifications must speak to each other. If proceeds are supposedly financing eligible expenditure, the government should be able to identify those amounts in execution systems, reconcile them to debt reports, and explain any substitution, delay, or reallocation (PEFA Secretariat, 2024).

Performance budgeting strengthens this chain when used carefully. OECD good practices on performance budgeting stress that performance information should inform decision-making without pretending to establish mechanical causality for every public outcome. That distinction matters in sustainable finance. It is reasonable to report on outputs, milestones, and selected outcome indicators. It is far less credible to imply that one bond issue "caused" a national climate result. Good systems distinguish attributable financing information from broader policy outcomes (OECD, 2019).

Internal control design should therefore test three things simultaneously: classification accuracy, financial traceability, and consistency of reported results. For debt offices and finance ministries, that often means building reconciliations among framework eligibility criteria, budget tags, debt allocation reports, and project reporting from line ministries. The more automated and auditable these links are, the stronger the assurance chain becomes.

Traceability is where sovereign sustainable finance usually becomes credible or fragile. A government should be able to move from a bond framework to a budget line, from a budget line to an executed expenditure, and from that expenditure to a public report explaining what was financed and on what basis it was deemed eligible. Where this chain is broken, second-party opinions and external reviews provide only partial comfort. They can support market confidence, but they do not replace the need for internal records that survive challenge. This is why repeat issuers such as Chile and Indonesia matter. Their public reporting cycles show how credibility depends not only on market access, but on the routine capacity to collect, reconcile, and explain information over multiple years (Chile Ministry of Finance, 2019; Ministry of Finance, Republic of Indonesia, 2023).

Internal audit sits closest to that routine control environment. The updated three lines model and the Global Internal Audit Standards are useful because they frame internal audit as an independent assurance function that should assess governance, risk management, and control design rather than own operational decisions. Applied to sovereign sustainable finance, that means testing whether framework governance is clear, whether source data are reliable, whether reporting processes are repeatable, and whether corrective actions are actually followed through. Internal audit does not decide which projects are green. Its job is to assess whether the system used to make and report those decisions is controlled, documented, and working as intended (The Institute of Internal Auditors, 2020, 2024).

Internal audit work programs should be risk based rather than ceremonial. Not every sovereign sustainable finance transaction requires a full assurance engagement every year, but issuers do need a rational basis for deciding where assurance effort is most needed. New frameworks, weak data interfaces, major methodology changes, or large concentrations of labeled spending are all obvious candidates. Sampling methods and control testing should then reflect those risks. This is where internal audit adds real value: by helping management understand which parts of the system are dependable and which are only assumed to be so (The Institute of Internal Auditors, 2024).

A useful discipline is to separate control assurance from policy endorsement. Internal auditors and SAIs should be able to examine whether a climate-related financing process is reliable without being drawn into political debate about the policy goal itself. That boundary protects independence. It also helps keep assurance focused on evidence, criteria, and performance information rather than on broad policy advocacy. The more visible sustainable finance becomes, the more important that professional boundary will be for credibility (INTOSAI, 2019a, 2019b; The Institute of Internal Auditors and INTOSAI, 2022).

Assurance plans can also be staged over time. A new framework may warrant heavier control testing in its first cycle, followed by more selective review once the process stabilizes. That staged approach avoids both extremes: under-scrutiny during the period of greatest uncertainty and over-audit once the process becomes routine.

4 The Role of Internal Audit and Supreme Audit Institutions

Internal audit is the first independent challenge inside the executive. The IIA's 2024 Global Internal Audit Standards define internal audit as a source of independent, risk-based assurance, advice, insight, and foresight that strengthens governance, decision-making, and credibility. In sovereign sustainable finance, that mandate translates into testing whether governance arrangements are clear, whether controls over proceeds and reporting operate as designed, whether guarantee and contingent liability processes are documented, and whether IT systems preserve audit trails (The Institute of Internal Auditors, 2024).

The IIA's Three Lines Model and the joint IIA-INTOSAI paper applying the model in the public sector are especially useful because they clarify that internal audit is not a substitute for management control, and SAIs are not substitutes for internal audit. Management owns the process. Internal audit tests it inside the executive. SAIs provide external public assurance. When those roles blur, accountability weakens. When they are clear, sovereign sustainable finance gains a layered and more credible oversight structure (The Institute of Internal Auditors, 2020; The Institute of Internal Auditors & INTOSAI, 2022).

SAIs add a broader accountability lens. INTOSAI's fundamental principles of public-sector auditing and its performance audit principles show why compliance alone is not enough in this area. A sustainable finance audit may legitimately ask not only whether the reported amounts reconcile, but whether the system delivers economy, efficiency, and effectiveness. That can include reviewing how project selection worked, whether controls reduced misclassification risk, whether reporting to investors matched reporting to parliament, and whether performance information was usable rather than ornamental (INTOSAI, 2019a, 2019b).

INTOSAI-P 12 makes the larger point explicit: SAIs create value by strengthening transparency, accountability, and trust in public institutions. Sustainable sovereign finance depends on exactly those conditions. A labeled bond can be sold without them, but it cannot build durable credibility without them. That is why SAI engagement should be viewed as a strategic asset, not as a compliance burden (INTOSAI, 2019c).

Supreme audit institutions add a broader layer because they can examine legality, regularity, performance, and the public value of the arrangements in question. INTOSAI principles matter here because sovereign sustainable finance is not only about whether spending matches a framework. It is also about whether governance arrangements improve transparency and whether public resources are being used effectively. Performance audit can therefore move beyond procedural compliance to ask harder questions: Were the selection criteria meaningful. Did reporting inform legislative scrutiny. Were outcomes measured in ways that support decision making. Those questions are especially important when sustainable borrowing is linked to broader policy claims about resilience and transition (INTOSAI, 2019a, 2019b, 2019c).

Public-sector coordination is easier when internal audit and SAIs are not working in isolation. Guidance developed jointly by the IIA and INTOSAI is helpful because it clarifies how the three lines model can operate in a public setting where responsibility is dispersed across ministries, agencies, and oversight bodies. In sovereign sustainable finance, the first line usually sits with operational managers in finance ministries and line ministries. The second line may include risk, compliance, or methodology functions. The third line is internal audit, while SAIs provide external public audit. Clarity about those boundaries helps avoid duplicated assurance on one issue and no assurance on another (The Institute of Internal Auditors and INTOSAI, 2022; The Institute of Internal Auditors, 2020).

SAIs can be particularly valuable when they connect financial and performance perspectives. A labeled borrowing program may comply with its framework and still tell legislators very little about whether the selected expenditures were timely, properly prioritized, or linked to stated results. Performance-oriented external audit can close that gap by asking whether the governance architecture actually helped public decision making. Those questions are demanding, but they are also where sovereign sustainable finance begins to justify itself as public finance rather than just market finance (INTOSAI, 2019b, 2019c).

External reviewers, by contrast, should be used with some humility. They can help assess framework alignment and may improve market confidence at issuance, but they usually work from narrower mandates than public auditors. Governments should therefore be careful not to present second-party opinions as if they were full substitutes for control assurance or performance audit. That distinction protects both the reviewers and the issuer. It keeps the market-oriented review in its proper place within a much wider public accountability structure (Bhutta et al., 2022; World Bank, 2022).

5 External Reviews, Performance Budgeting, and Public Credibility

External reviews such as second-party opinions, framework assessments, and post-issuance verification have a clear place, but that place should be understood correctly. They are market-facing assurance devices. They can help investors interpret whether a framework aligns with recognized principles and whether reported use of proceeds follows stated rules. They do not replace the government's own control systems, and they do not replace public audit. The danger begins when governments treat external opinions as a substitute for internal traceability or legislative accountability (Di Marco et al., 2023; World Bank, 2022).

Performance budgeting provides a bridge between these worlds because it helps governments report not only what was financed, but why it was financed and how results will be monitored. OECD guidance is helpful precisely because it is pragmatic. Performance information should complement fiscal and control information, not overwhelm it. For sovereign sustainable finance, that means linking financing to a limited set of decision-relevant indicators and reporting cycles that can actually be maintained over time (OECD, 2019, 2024).

Country practice shows why substance matters. Chile and Indonesia have both supported sovereign thematic issuance with allocation and impact reporting that is more useful because it sits within stronger public finance routines rather than being produced as a one-off marketing document. Nigeria's experience is also instructive because it demonstrates how framework ambition must be matched by reporting capacity if credibility is to deepen across successive issues (Chile Ministry of Finance, 2019; Debt Management Office Nigeria, 2024; Ministry of Finance, Republic of Indonesia, 2023).

The mature assurance model is therefore cumulative. Management systems create traceability. Internal audit tests control effectiveness. SAIs provide public assurance and performance scrutiny. External reviewers help market interpretation. Performance budgeting links financing to policy logic without overstating attribution. When these pieces reinforce one another, sovereign sustainable finance can survive skepticism and scale with credibility. When they do not, the framework may still look polished, but it will remain fragile.

External reviews still have a role, but their value depends on how they connect to public systems. Second-party opinions, verification reports, and external framework assessments can help standardize market communication and reduce informational uncertainty for investors. Yet they do not substitute for audit-ready internal evidence or for legislative accountability. The point is not to choose between market assurance and public assurance. It is to connect them. A strong sovereign issuer can explain how external reviewers fit alongside internal controls, internal audit, and external public audit rather than allowing those assurance streams to operate as separate worlds (Bhutta et al., 2022; World Bank, 2022).

Performance budgeting adds another useful layer because it forces the discussion toward results as well as compliance. OECD good practices on performance budgeting suggest that linking resources to objectives can strengthen accountability when the indicators are clear, decision relevant, and not overly proliferated. In sustainable sovereign finance, that means resisting the temptation to report dozens of loosely connected metrics. A smaller set of meaningful indicators that can be traced to financing decisions and implementation realities is usually more valuable. Strong assurance is not about producing more documents. It is about producing evidence that is usable for scrutiny and credible enough to withstand it (OECD, 2019, 2024).

Public communication after assurance work also deserves more care than it often receives. Audit and review findings should not disappear into technical annexes. Where control weaknesses are identified, governments should explain what will be corrected and when. That follow-up discipline is part of what turns assurance into a credibility asset rather than a one-off compliance event.

There is a sequencing benefit here as well. When control expectations are set early, reporting becomes easier, audit becomes more efficient, and policy debate can focus less on data disputes and more on substance. That is one reason strong assurance improves not only accountability but also the usability of sovereign sustainable finance for decision makers.

6 Conclusion

Sovereign sustainable finance depends on more than good intentions and well-designed frameworks. It depends on the ability of the state to prove that funds were allocated according to rule, that claims about climate or social purpose are traceable, and that reporting to investors is consistent with reporting to domestic oversight institutions. Assurance is the mechanism that makes those claims believable.

The practical lesson is clear. Internal control, internal audit, supreme audit institutions, and performance budgeting should be built into sustainable finance design from the outset rather than added after issuance. External reviewers can support market confidence, but they cannot replace public accountability. The sovereigns that build the strongest assurance chains will not only reduce reputational risk. They will also improve the quality of financing decisions themselves.

What public credibility requires, then, is an assurance chain that is continuous from framework design to audit follow-up. Before issuance, control owners and evidence requirements should be clear. During execution, proceeds and tags should be traceable through budget and accounting systems. After issuance, internal audit, external reviewers, and where relevant SAIs should be able to test the claims being made without reconstructing the evidence base from scratch. That continuity is demanding, but it is also what distinguishes mature sovereign sustainable finance from disclosure that looks impressive but cannot be fully defended later (PEFA INTOSAI, 2019a; Secretariat, 2024; The Institute of Internal Auditors, 2024).

In the end, assurance is less about adding a final seal of approval than about protecting the integrity of the whole financing architecture. When that integrity is visible, governments gain more than market access. They build trust with legislatures, auditors, and citizens as well. That trust is slow to build and easy to lose, which is why strong oversight remains one of the least optional parts of sovereign sustainable finance.

Performance budgeting also works better when the indicator set is disciplined. Too many indicators create reporting noise and encourage weak data collection. Too few can make the program look uninformative. The more effective balance is usually a small group of indicators with clear definitions, stable methods, and strong relevance to policy choices. That is the kind of evidence legislators and auditors can actually use when asking whether financing claims and public outcomes line up.

The most persuasive sovereign issuers will therefore be those that can show how controls, audit, and performance information fit together rather than presenting them as separate compliance products. Once those links are visible, sustainable finance stops looking like a public-relations layer and starts to look like a governed part of the state's fiscal system.

A strong evidence repository can help enormously here. When documents, reconciliations, methodology notes, and control records are maintained systematically, both internal and external reviewers can work more efficiently and with less disruption to operating teams.

This may sound administrative, yet it sits very close to the core of public trust. Sustainable finance claims are easiest to defend when the evidence behind them has been organized as if scrutiny were expected from the start.

For governments, this means assurance should be budgeted and staffed like a core operating requirement rather than treated as a peripheral communications cost. The more visible sustainable borrowing becomes, the less viable it is to rely on improvised evidence collection or last-minute reconciliation.

This final point brings the chapter back to its starting claim. Assurance is not a wrapper around sovereign sustainable finance. It is one of the mechanisms that determines whether the financing itself will remain credible over time.

That expectation of continuous scrutiny is ultimately healthy. It encourages governments to build sustainable-finance systems that are not only visible, but also defensible when evidence is requested in detail.

References

Bhutta, U. S., Tariq, A., Farrukh, M., Raza, A., & Iqbal, M. K. (2022). Green bonds for sustainable development: Review of literature on development and impact of green bonds. *Technological Forecasting and Social Change, 175*, Article 121378. https://doi.org/10.1016/j.techfore.2021.121378

Chile Ministry of Finance. (2019). *2019 report: Allocation and environmental impact (green bonds).* Ministry of Finance, Government of Chile.

Debt Management Office Nigeria. (2024). *Green bond*. Federal Government of Nigeria.

Di Marco, R., Dong, T., Malatincová, R., Reuter, M., & Strömsten, T. (2023). Symbol or substance? Scrutinizing the risk transparency premise in marketized sustainable finance: The case of TCFD reporting. *Business Strategy and the Environment, 32*(6), 3027–3052. https://doi.org/10.1002/bse.3285

INTOSAI. (2019a). *ISSAI 100: Fundamental principles of public-sector auditing*. INTOSAI.

INTOSAI. (2019b). *ISSAI 300: Performance audit principles*. INTOSAI.

INTOSAI. (2019c). *INTOSAI-P 12: The value and benefits of supreme audit institutions: Making a difference to the lives of citizens*. INTOSAI.

Ministry of Finance, Republic of Indonesia. (2023). *Green sukuk allocation and impact report 2023.* Ministry of Finance, Republic of Indonesia.

OECD. (2019). OECD good practices for performance budgeting. *OECD Publishing*. https://doi.org/10.1787/c90b0305-en

OECD. (2021). Green budget tagging: Introductory guidance and principles. *OECD Publishing*. https://doi.org/10.1787/fe7bfcc4-en

OECD. (2024). Green budgeting in OECD countries 2024. *OECD Publishing*. https://doi.org/10.1787/9aea61f0-en

PEFA Secretariat. (2024). *Supplementary framework for assessing climate-responsive public financial management*. Public Expenditure and Financial Accountability Secretariat.

The Institute of Internal Auditors, & INTOSAI. (2022). *Applying the three lines model in the public sector*. The Institute of Internal Auditors and INTOSAI.

The Institute of Internal Auditors. (2020). *The IIA's three lines model: An update of the three lines of defense*. The Institute of Internal Auditors.

The Institute of Internal Auditors. (2024). *Global internal audit standards*. The Institute of Internal Auditors.

World Bank. (2022). *Sovereign green, social and sustainability bonds: Unlocking the potential for emerging markets and developing economies*. World Bank.

GPSR Compliance
The European Union's (EU) General Product Safety Regulation (GPSR) is a set of rules that requires consumer products to be safe and our obligations to ensure this.

If you have any concerns about our products, you can contact us on

ProductSafety@springernature.com

In case Publisher is established outside the EU, the EU authorized representative is:

Springer Nature Customer Service Center GmbH
Europaplatz 3
69115 Heidelberg, Germany

www.ingramcontent.com/pod-product-compliance
Ingram Content Group UK Ltd.
Pitfield, Milton Keynes, MK11 3LW, UK
UKHW020936180926
475692UK00003B/73
9783032332387